JUL **2 3** 2010

D0161426

STARS AND GALAXIES

Greenwood Guides to the Universe
Timothy F. Slater and Lauren V. Jones, Series Editors

Astronomy and Culture
Edith W. Hetherington and Norriss S. Hetherington

The Sun
David Alexander

Inner Planets
Jennifer A. Grier and Andrew S. Rivkin

Outer Planets
Glenn F. Chaple

Asteroids, Comets, and Dwarf Planets
Andrew S. Rivkin

Stars and Galaxies
Lauren V. Jones

Cosmology and the Evolution of the Universe
Martin Ratcliffe

STARS AND GALAXIES

Lauren V. Jones

Greenwood Guides to the Universe
Timothy F. Slater and Lauren V. Jones, Series Editors

GREENWOOD PRESS
An Imprint of ABC-CLIO, LLC

Santa Barbara, California • Denver, Colorado • Oxford, England

Library of Congress Cataloging-in-Publication Data

Jones, Lauren V.
   Stars and galaxies / Lauren V. Jones.
      p. cm. — (Greenwood guides to the universe ; v. 6)
   Includes bibliographical references and index.
   ISBN 978-0-313-34075-8 (hard copy : alk. paper) — ISBN 978-1-57356-749-7 (ebook)
1. Stars—Popular works. 2. Galaxies—Popular works. I. Title.
   QB801.6.J66 2010
   523.8—dc22   2009034909

14 13 12 11 10   1 2 3 4 5

This book is also available on the World Wide Web as an eBook.
Visit www.abc-clio.com for details.

ABC-CLIO, LLC
130 Cremona Drive, P.O. Box 1911
Santa Barbara, California 93116-1911

This book is printed on acid-free paper ∞

Manufactured in the United States of America

To my husband, Tarek,
and my children, Omar Zakaryah and Leilah Sorayah.
I love you more than the whole universe!

Contents

Series Foreword

Not since the 1960s Apollo-era has the subject of astronomy so readily captured our interest and imagination. In just the past few decades, the a constellation of space telescopes, including the Hubble Space Telescope, have peered deep into the farthest reaches of the universe and discovered supermassive black holes residing in the center of galaxies. In concert, giant telescopes spread around the globe on Earth's highest mountaintops have spied planet-like objects larger than Pluto lurking at the very edges of our solar system and carefully measured the expansion rate of our universe. Simultaneously, meteorites from Mars with bacteria-like fossil structures have spurred repeated missions to Mars with the ultimate goal of sending humans to the red planet. Given that astronomers have recently discovered hundreds of planets beyond our solar system, we are given pause and a rationale to capture what we now understand about the cosmos in these Greenwood Guides to the Universe volumes as we prepare ourselves to peer even deeper into the universe's secrets.

As a discipline, astronomy covers a range of topics stretching from the central core of our own planet outward past the Sun and nearby stars to the most distant galaxies of our universe. As such, this set of eight volumes systematically does the same, covering the main components of our solar system (*The Sun; Inner Planets; Outer Planets;* and *Asteroids, Comets, and Dwarf Planets*) and all the major structures and unifying themes of our evolving universe (*Stars and Galaxies; Cosmology and the Evolution of the Universe;* and *Astronomy and Culture*). Each volume comprises a narrative discussion highlighting the most important ideas of major celestial objects and how astronomers have come to understand their nature and evolution. In addition to describing astronomers' most current understandings of the topics covered, these volumes also include perspectives on the historical and pre-modern understandings that have motivated us to pursue deeper knowledge.

The ideas presented in these Greenwood Guides to the Universe volumes have been meticulously researched and carefully written by experts to provide students and interested non-expert readers with the most current and scientifically accurate information and understandings of astronomers working today. Some astronomical phenomena we just do not understand very well, and the volume authors have tried to distinguish between theories that have wide consensus and those that are as yet unconfirmed. Because astronomy is a rapidly advancing science, some of the concepts presented in these pages will almost certainly become obsolete as advances in technology yield previously unknown information. If unintentional errors exist in these volumes, they are our responsibility as series editors. Astronomers' share and value a worldview that scientific knowledge is subject to change as the scientific enterprise makes new and better observations of our universe. Our understanding of the cosmos evolves over time, just as the universe evolves, and what we learn tomorrow depends on the insightful efforts of dedicated scientists from yesterday and today. We hope these volumes reflect the deep respect we have for the scholars who have worked, are working, and will work diligently in the public service to uncover the secrets of the universe.

Lauren V. Jones, Ph.D.
Timothy F. Slater, Ph.D.
Series Editors

Preface

Astronomy is, literally, the systemized knowledge of stars. Early astronomy was just that—the mapping of the known fixed heavenly bodies so that new or moving ones could be easily identified. Modern astronomy embodies the knowledge of every part of the cosmos, including the study of galaxies and the universe itself. The volumes in the Greenwood Guides to the Universe Series attempt to provide an overview of many of the key areas of study in this field, written at a level appropriate for high school students, undergraduates, and interested public library patrons.

The first four chapters of *Stars and Galaxies* discuss the nature and properties of stars. Chapter 1 delves into the historical and scientific definition of a star and the physical properties of stars. Chapter 2 explores the properties of stars in the context of the most important tool to astronomers, the H-R Diagram. Chapter 3 tells the story of how stellar evolution was derived from simple observations of stars. Chapter 4 describes all the categories of variable stars and talks about star systems.

Chapters 5–10 transition to a discussion of large systems of stars called galaxies. Chapter 5 describes the definition and physical characteristics of a galaxy. Chapter 6 is devoted to the Milky Way, the galaxy in which the Earth is located; it describes what astronomers know about the Milky Way and how they know it. Chapter 7 discusses arms in disk galaxies, introducing such topics as spiral density waves, rotation curves, and dark matter.

Galaxy interactions are the main topic of chapter 8, which also explores the relationship between interactions and the formation of bars in disk galaxies, a discussion that leads into the next chapter's discussion of active galaxies. Chapter 9 introduces the four main types of active galaxies along with the unified model of the central engine, which powers these objects. Chapter 10 is about clusters and groups of galaxies as well as galaxy evolution. *Stars and Galaxies* tells the story of what we currently know about both of those celestial objects as well as how we know it. Readers of this volume will get a deep understanding of how astronomers have studied stars and galaxies to learn about their physical nature and how stars and galaxies evolve over timescales longer than human existence.

The text of most chapters is augmented by informative sidebars and illustrations, as well as profiles of astronomers who have furthered our knowledge of stars and galaxies. Words that appear in boldface type are defined in a glossary of key terms. Each chapter concludes with a list of print or electronic resources for further information and updates. Also included are a general bibliography and a detailed subject index.

Acknowledgments

The author would like to thank her family for their support throughout the long process of writing this volume. Also, the author would like to thank those who taught her the most about astronomy and how science works (in chronological order): Dr. Profeta, Mr. Sweeney, Debbie Elmegreen, Henry Albers, Cindy Schwarz, Fred Chromey, Mort Tavel, Emilia Belserene, Alan Harris, John Gaustad, Anatoly Vladimirovich Zasov, Nikolai Nikolaievich Shakura, Mikhail Sazhin, Bill Keel, Henry Emile Kandrup, Richard Joseph Elston, Elizabeth Lada, Gus Muench, Tim Spahr, Karl Haisch, Tim Slater, Larry Marschall, Ed Prather, and Esther Hopkins. Each of these individuals contributed significantly to developing my ability to write this volume.

Finally, to those who served as role models for the author, she would like to express her gratitude for their leadership and exemplary scientific integrity here: Vera Rubin, Saul Perlmutter, Robert Kennicutt, Robert Kirschner, John Huchra, and Dorritt Hoffleit.

Introduction

The study of stars probably began before recorded human history. From an anthropological point of view, humans probably used the stars and their regular patterns to predict seasonal changes. Archeoastronomy is the modern science of searching for evidence of this, and archeoastronomers have found significant evidence that ancient cultures monitored the motion of the stars, sun, and moon for various purposes. The study of galaxies is still in its infancy. Astronomers were not even aware that galaxies existed until the early part of the 20th century. Nonetheless, much has changed since that realization. Astronomers have learned a lot in a very short period of time.

Modern astronomy is an entirely different endeavor than astronomy in any other time period. The only thing modern and ancient astronomers have in common is their subjects. Ancient astronomers, of course, did not have the tools we use today to gather the kind of data we now have about stars and galaxies and their physical natures. Ancient astronomers were responsible for mapping the positions of the stars and noticing that some moved relative to others, distinguishing them from the other stars as "wanderers," or planets.

Early modern astronomers built on that knowledge and used telescopes to find other kinds of objects. They noted all the "fuzzy" objects, or nebulae, that they could find and recorded their positions. They began to be able to see the surface and cloud features of the planets and could distinguish the shapes of some nebulae as spiral. For the pursuit of the study of stars, the most important discovery in early modern astronomy was that of clusters of stars.

In the early 20th century, the telescope was the main tool used by astronomers; however, by the 1920s, the spectroscope was widely used. The spectroscope allowed modern astronomers to study more than just points of light in the night sky. Now astronomers could study the light itself. This is how astronomers were able to begin the discovery of the physical nature of stars and the existence of the "island universes" we now call galaxies.

Because of the ability to collect data on photographic plates, astronomers were able to collect more data than they could analyze. Suddenly, there were more data than there were astronomers to analyze the data. This marked the advent of women entering the field of astronomy. Many women were educated in the early 20th century, but there were few places an educated woman could work. Astronomy was in need of educated workers, and there were plenty of women who wanted to be involved in astronomy.

In the United States, Vassar Female College (now known as Vassar College) had one of the most productive astronomy programs, led by the first American female astronomer, Mariah Mitchell. Many women graduated from her program at Vassar College and went on to work at the Harvard Observatory as "computers." Their contribution to the field of astronomy is enormous and somewhat untold. Many have heard of these human computers, the most famous of whom is Annie Cannon, but few know of their many accomplishments.

As a result of this frenzy of analyzing astronomical data from stars along with the work of Ejnar Hertzsprung and Henry N. Russell, astronomers pieced together stellar evolution. Many of the details of stellar evolution are still unknown, but the big picture is fairly clear. More technological advances have contributed to the detail we now possess.

The 1950s saw the advent of radio astronomy and the beginning of digitized data. Soon after this came infrared astronomy, ultraviolet astronomy, x-ray astronomy, charge coupled devices (CCDs, or digital cameras), and much more. Digitized data meant that more data could be accumulated and the storage of it would take far less space. Digitized data is also more objective, and the advent of easily attainable digital data revolutionized the study of astronomy. No longer did astronomers rely on a person's ability to "flyspackle" (compare the sizes of silver oxide spots on a photographic plate). Digital data could be analyzed objectively in an automated way by modern computers (not people, this time!).

Using almost the entire electromagnetic spectrum to study objects in the sky, astronomers now know much more about the physical nature of stars and galaxies. Using data taken over the last 100 years, astronomers have more insight into some of the details of stellar evolution and are beginning to understand the principle role of merger events in galaxy evolution. Using telescopes all over the globe and in outer space, astronomers can see more clearly (literally!) what a star or a galaxy is and how it behaves over short time periods.

Modern astronomers who study stars ask questions like the following: How does dust form on the surface of carbon stellar remnants? How does the T Tauri phase affect the formation of planets and asteroids in a distant solar system of planets? What is the maximum mass a star can have before it blows itself apart? What is the minimum mass required for a remnant to become a black hole instead of a neutron star?

Modern astronomers who study galaxies ask questions like the following: What does a galaxy's shape tell us about its physical nature? How much can a galaxy change because of interactions with other galaxies? What is the nature of the central engine that powers active galactic nuclei? What can clusters of galaxies tell us about galaxy evolution?

Despite almost a century of modern technology helping astronomers study stars and galaxies, there are still many questions to be answered. The study of stars and galaxies is an exciting part of astronomy, and it is, perhaps, one of the longest-lived areas of modern astronomy.

1

What Is a Star?

Pumbaa: Ever wonder what those sparkly dots are up there?
Timon: Pumbaa. I don't wonder; I know.
Pumbaa: Oh. What are they?
Timon: They're fireflies. Fireflies that uh…got stuck up in that big…bluish-black…thing.
Pumbaa: Oh. Gee. I always thought that they were balls of gas burning billions of miles away.
Timon: Pumbaa, wit' you, everything's gas.

From *The Lion King*

Many have wondered, as these characters from the movie *The Lion King* have, what the stars are. One way to try to determine what stars are is by watching them move. Anyone can do this experiment. Just draw a picture of what you see when you go out on a clear night. Be sure to include any objects on your horizon as reference points. Then wait about an hour or more and go back out to look at the sky. Check your drawing against what you see.

THE MOTIONS OF THE STARS

Because of the rotation of Earth on its axis, the stars appear to rise in the east, move high in the sky (although not necessarily to a point directly overhead), and then set in the west. This is true, no matter where you are in the world. (Although, east and west are difficult to define at the north and south poles.) In the northern hemisphere, stars rise at an angle to the eastern horizon. So, stars appear to rise high in the southern sky (not to a point overhead).

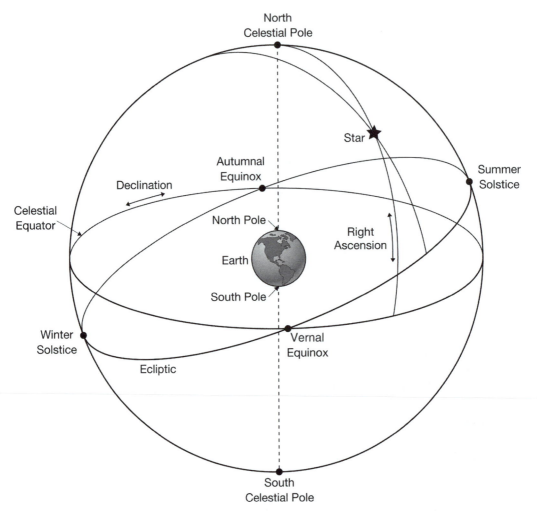

Figure 1.1 This is a diagram of a celestial sphere. A celestial sphere is a tool used in astronomy to represent the motions of objects in the night sky due to Earth's rotation on its axis. In this model, Earth is a stationary point in the center while the Sun and stars appear to move around it. This is consistent with what we observe on the surface of Earth, but not representative of the actual motions of bodies. From Earth, in reality, the Sun and star only *appear* to move because Earth is rotating on its axis. [Jeff Dixon]

In the southern hemisphere, the stars rise in the east as they do everywhere else on Earth, but before it sets in the west, a star will rise to a point high in the *northern* sky (not to a point overhead). It is disorienting to someone who has only seen the northern hemisphere sky to observe the sky in the southern hemisphere (and vice versa, of course). Even to those who do not usually pay very close attention to the sky, something seems "not right."

The only place on Earth where a star can rise due east and pass through the point in the sky directly overhead before setting due west is at the equator. At this point on the surface of Earth, a star that rises due east will follow a path through the sky will take it through the zenith (otherwise known as the point in the sky directly overhead).

The stars appear to move all together. One can divide the stars into groups (called **constellations**) that maintain their shape and relative positions throughout the night. This is because the stars are *not* actually moving. The stars only *appear* to move because Earth is rotating on its axis. In fact, the motions of most stars relative to the Sun are so small that it takes thousands of years for almost any visible change to manifest itself. The motions of stars relative to the Sun are so small because the stars are so far away.

Constellations

Constellations have been defined and described by many different cultures in the history of human-kind. The constellations used by astronomers today were, for the most part, designated by the ancient Greeks. There are 88 constellations defined by the International Astronomical Union. Those in the northern hemisphere are Greek in origin, but in the southern hemisphere some constellations were given their boundaries and names by the early explorers. Most of the Greek constellations have myths associated with them. These stories tell of the creatures or characters they represent and how they ended up in the night sky (for example, see Ovid's *Metamorphosis* which describes the mythological origin of most of the Greek constellations). While other cultures have grouped the stars into different constellations, it is the Greek names that were adopted by modern astronomers. Interestingly, the constellation Orion (the Hunter, in Greek mythology) is similar in many different cultures. It has similar boundaries, as well as a similar form. Many cultures (including, Chinese, Egyptian, Native American cultures) designated the stars in this region of the sky as representing a human man.

Some of the so-called constellations with which we are most familiar are not constellations at all. For example the Big Dipper is not a constellation. It is an **asterism,** or group of stars that are part of one or many constellations. In the case of the Big Dipper, the stars that make up the asterism are part of the constellation Ursa Major. Another well-known asterism is the Summer Triangle. Stars that make up the Summer Triangle are from three different constellations: Lyra, Cygnus, and Aquila.

Studying the relative motions of stars merely tells us that they are very far away. Nearly all stars stay in the same place relative to one another night after night, month after month, year after year. Very few stars appear to move relative to the others. Stars that move relative to others are called high **proper motion** stars. These stars are very close to our Sun, compared to the others, and are moving exceptionally fast—relative to our Sun's motion through space.

Several objects look like stars and also appear to move relative to the nearly constant stellar background; these are the planets. The word "planet" comes from the Greek word for "wanderer." The planets were so named *because* they appeared to move relative to the other stars. At that time, the Greeks did not know that the planets and the stars were fundamentally different objects; they were merely noting the different motions of the two objects. The motions of the planets are so significant that one can observe changes in positions of the three nearest planets (Mercury, Venus, and Mars) on at least a weekly scale. That is, one might not notice changes from night to night, but if one looks at a planet's position relative to the stars this week and then looks again next week, one will probably be able to detect that the planet has changed position with respect to the stars.

The positions of objects in the night sky are designated by coordinates called **right ascension** and **declination.** Right ascension is like longitude on Earth; it is a measure of how far a star is behind the constellation Aries, in units of hours, minutes, and seconds of **sidereal time.** The constellation Aries marks the "prime meridian" of the sky. On Earth the prime meridian is the line that goes through Greenwich, England. In this location on Earth, there was an observatory, and time on Earth was kept by this observatory for centuries, so it became the reference line to the lines of longitude on the globe. In the sky, the path of the Sun relative to the stars crosses the celestial equator (a projection of Earth's equator on the sky) in two places: the Vernal Equinox and the Autumnal Equinox. Our calendar year used to begin on the Vernal Equinox, so it was chosen to be the reference line on the celestial sphere.

The time it takes for Earth to rotate once on its axis relative to the stars is about 23 hours 56 minutes. This is less than what we call a "day" on Earth. The reason that it takes less time for Earth to rotate on its axis relative to the stars than it does to rotate on its axis relative to the Sun is because of its orbit around the Sun. As Earth rotates on its axis it moves in a nearly circular orbit around the Sun. This means that Earth must make a little more than one full rotation to get the Sun to the same relative position in the sky. So, the solar day is longer than the sidereal day by about 4 minutes.

Thousands of years ago, people who studied the stars and recorded their positions were called astrologers. Nowadays, astrologers still track the positions of the planets relative to the stars. Then and now, astrologers believe that the relative positions of stars and planets have some meaning in the lives of humans on Earth. Although science has found no evidence for such a connection, modern astronomers often use the data recorded by ancient astrologers to learn more about the objects astrologers watched.

One thing the ancient astrologers did exceptionally well was to carefully note the times and positions of irregular phenomena. It is because of ancient astrologers' records that the periodic nature of comets was discovered. It is also ancient astrologers' records (in combination with modern astronomical studies) that tell us how the Crab Nebula came to be. Ancient astrologers were the first record keepers and the first data gatherers in the field now known as astronomy. While modern astrologers do not contribute to modern astronomy, there is an interesting overlap between the two practices thousands of years ago, before science truly existed.

Modern astronomers study the stars, and other objects in space, to better understand how and why they are what they are. Astrologers are not interested in *what* stars and planets are, but *where* they are in the sky; they believe that this has some relevance to human lives. There is, however, no scientific evidence that the relative positions of planets and stars have *any* influence on human lives.

So what *is* a star? Studying stars' relative motions only told us that the stars are very far away. Stars do not move significantly, for the most part, relative to one another. A human, or even a generation of humans, will not detect any relative motions of most stars without the aid of modern instruments of astronomy. The motions of the stars relative to one another, as well as their apparent motions relative to Earth, did not tell us much about what stars are. Until the birth of modern astronomy, there was no way to answer the question of what a star is beyond mere speculation.

THE SUN IS A STAR

The closest star to Earth is the Sun. The details of what we know about the Sun can be found in *The Sun,* another volume in the Greenwood Guides to the Universe series. To sum up, the Sun is made up mostly of hydrogen gas, with a significant portion of the Sun made of helium gas. A small percentage of the Sun contains other elements in gas form. (So, Pumbaa was right. Stars *are* giant balls of gas.) How we know what the Sun and other stars are made of will be discussed in much more detail later in this chapter.

Even though the Sun is made entirely of gas, it is very dense at its core. The Sun is not a solid body, like a planet, so even if it weren't too hot, a person couldn't stand on its surface, because it doesn't have a solid surface like the surface of Earth. However, the density of the Sun near the core is very high, much higher than the density of the Earth, even at *its* core. That extremely high density is necessary for the Sun to produce light.

The Sun produces energy through the process of **thermonuclear fusion** of hydrogen into helium in its core. This means, essentially, that hydrogen atoms are being smashed together producing energy, as well as heavier, helium atoms. Thermonuclear fusion will be discussed in much more detail in chapter 3. After the energy has completed its long journey from the center of the Sun to its edge, then through space, this energy is received by us on Earth in the form of **electromagnetic radiation** (or, light).

Our Sun is huge, compared to Earth. The Sun has a mass of about 2×10^{30} kg, which is about 330,000 times more massive than the Earth. The Sun has a diameter of almost 1.4 million km. This is about 110 times the diameter of Earth. If you compare the average density of the Sun to the average density of Earth, you would find that the Sun has an average density that is about 1/3 that of Earth's. Although the core of the Sun has a much greater density than the core of Earth, the vast majority of the Sun has a much lower density than even Earth's atmosphere.

So, much is already known about the Sun. We have a sense of its size and scale, as well as what it is and how it works. Still, there are some lingering questions: Are most stars like the Sun? (Most planets are not like Earth.) What are the ranges of size, mass, brightness, and surface temperature that stars can have? Do all stars have the same mechanism for producing energy in their cores? Do

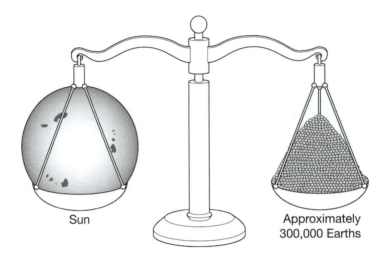

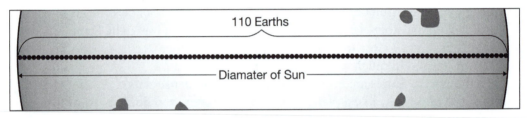

Figure 1.2 The diagram on the left shows the relative mass of the Sun to Earth. The Sun is about 300,000 times more massive than Earth. The diagram on the right shows the relative sizes of the Sun and Earth. The Sun is so large that about 110 Earths could fit across its diameter. [Jeff Dixon]

stars ever "die" or cease to exist? If so, how are stars "born" or formed? These and other questions will be answered in this chapter and in this volume.

STARS COME IN MANY SIZES

It turns out that quite a lot of stars *are* like the Sun; however, there are many stars that have different sizes, masses, brightnesses, or surface temperatures. When astronomers talk about other stars, they always use the Sun as a reference. For example, the mass of a star is usually given in **solar masses.** A solar mass is the mass of the Sun (1.989×10^{30} kg). The symbol for a solar mass is $M_\odot$. The **solar radius** ($R_\odot$) is 6.96×10^8 m. The brightness of stars is usually discussed in terms of **luminosity,** which is a measure of the total light emitted by a star, rather than the light received from a star. The **solar luminosity** ($L_\odot$) is 3.827×10^{33} erg/sec or 3.827×10^{26} Watts (that's *much* brighter than the typical lightbulb!). The surface temperature of the Sun is 5,500 K. (Here, the K is for Kelvin, a measure of temperature that is similar to the Celsius scale.) Surface temperature is one of the few characteristics of stars that are never used in reference to our Sun.

Kelvin Temperature Scale

The Kelvin temperature scale was developed by William Thomson in 1848. Thomson was Scottish physicist at the time, educated at Glasgow University and Cambridge University. Thomson studied thermodynamics and the theory of heat. Specifically, Thomson studied Carnot's theory of heat as laid out by the Carnot cycle. Carnot was interested in determining the most efficient way to use heat energy to do work. His studies showed the importance of the second law of thermodynamics in resolving this question. The second law of thermodynamics simply states that there is a limit to the amount of heat that can be used to do work. Carnot's theory of heat defined the efficiency of an engine that uses heat to do work by the difference between the temperatures of the hot and cold reservoirs within the engine divided by the temperature of the hot reservoir.

Frustrated that the definition of temperature was derived from the workings of a gas thermometer, Thomson proposed a temperature scale for which 0 degrees corresponds to absolute zero, so that the scale would be "independent of the physical properties of any specific substance." Absolute zero is the temperature at which molecular energy is at a minimum, so that an object with a temperature of 0 could not transfer heat. Below absolute zero, temperatures do not exist. The proposed temperature scale was not accepted immediately and was eventually modified somewhat. In the end it was adapted in its current form. The temperature differences are the same as the Celsius scale, and the freezing and boiling points of water are 273 K and 373 K, respectively.

Essentially, the Kelvin scale is the Celsius scale shifted by 273.15 degrees so that the temperature at which water freezes is 273 K, or 0° C. The name of the Kelvin scale is from William Thomson's title, Baron Kelvin of Largs, given to him by the British government in 1892. Largs is the burgh in which Thomson resided, and Kelvin is the name of the second largest river in Glasgow.

The theoretical lower limit for a star's mass is 0.08 $M_\odot$. If the mass of the star is lower than 0.08 $M_\odot$, its core should not be dense enough or hot enough to fuse hydrogen. Since it will not be able to fuse hydrogen in its core, it will not be able to produce energy or the force necessary to balance the star and prevent gravitational collapse. This lower limit is a calculated limit based on what astronomers know about stars and how they work. There exists an object, currently classified as a main sequence star (fusing hydrogen in its core) that has a mass of 0.04 $M_\odot$.

The theoretical upper limit for a star's mass is as yet unknown. The current accepted limit is anywhere from 100–120 $M_\odot$. Astronomers hypothesize that there must be a physical upper limit since the existence of a star is entirely a balancing act between two forces. At some point, it is predicted that the outward forces (balancing gravity) should be too strong and cause the star to blow apart. The highest mass star ever observed (HD 92350) has a luminosity corresponding to a mass of about 120 $M_\odot$.

Star Names

Star names are given in one of three ways. Most stars have at least one name, corresponding to the catalog in which it appears. If the star is bright enough to be seen with a naked eye, it may have two names (one from a catalog and one from the constellation in which it is located). Very bright stars usually have proper names also.

The brightest stars in the northern hemisphere sky were named by the ancient Arabian, Greek, or Roman astronomers. The vast majority of the names with which many of us are familiar are Arabic (like Vega, Betelguese, Rigel, Altair, and Aldebaran). These names were given to the stars when astronomers of Arabia adopted the Greek constellations, but gave their own names to the stars within them. We know this because, generally, the Arabic name has a meaning that is relevant to the Greek constellation. Some stars, however, have retained their Greek or Latin names (like Polaris, Sirius, Capella, Castor, and Pollux). There are at least several hundred stars with proper names like these.

In addition to proper names, all the stars one can see with the naked eye (and some that require a telescope to view) have names that are determined by the constellation in which it resides and the relative brightness of the star within its constellation. These names are known as a star's Bayer designation. For example, Sirius is the brightest star in the constellation Canis Major, so the Bayer designation for this star is Alpha Canis Majoris. Since "alpha" is the first letter in the Greek alphabet, the use of "alpha" in the Bayer designation indicates that the star in question is the brightest star in its constellation. The constellation name is changed slightly. The original name is Canis Major, but in the Bayer designation it is Canis Majoris. The change is to the possessive case in Latin. This means that a literal translation of the words "Canis Majoris" is "of Canis Major." The Bayer designation for Sirius is, then, the brightest star in Canis Major, or Alpha Canis Majoris.

If the star is not the brightest star in the constellation, the letter of the Greek alphabet will be different. For example, Bellatrix is the third brightest star in the constellation Orion. So, its Bayer designation would be Gamma Orionis. "Gamma" is the third letter in the Greek alphabet. For constellations that contain more visible stars than the Greek alphabet will allow, modern astronomers have extended the Bayer designation by using the English alphabet and a double letter designation in English to include even fainter stars.

Finally, the vast majority of stars that are not easily seen with the naked eye have names that are related to a catalog where they are listed. There are many catalogs of stars. The most well-known catalog that lists faint stars is the Bonner Durchmusterung catalog. This catalog was compiled in the 19th century. Stars in this catalog have names that look like this BD+38° 3238. This is actually the BD number for Vega. In the Bonner Durchmusterung, Vega is the 3,238th star in the declination range between 38 and 39 degrees north (positive). Another commonly used catalog is the Henry Draper Memorial Catalog. This catalog lists stars by right ascension, rather than declination, but assigns each star a number. In the Henry Draper Catalog, Vega is HD 172167. There are many catalogs that list stars, and some stars (like Vega) have as many as 40 names.

• •

Surprisingly, the smallest sized stars are *not* the least massive stars. The smallest known stars are compact remnants of very large stars. The smallest stars are called **neutron stars.** These stars no longer fuse anything in their cores and are no longer made of hydrogen and helium. Neutron stars can be as small as 10 km across. That is about 140,000 times smaller than our Sun.

The largest stars are also nothing like our Sun in terms of their composition or their method of producing energy. These stars are known as **supergiants.** The largest known supergiant star (R Cassiopeiae) has a radius 1,800 $R_{\odot}$.

BRIGHTNESS OF STARS

The brightest known stars have luminosities up to 100,000 times the solar luminosity. These stars are supergiant stars: they are much larger (and can be much hotter) than our Sun in addition to being much more luminous. Also, supergiant stars produce energy using thermonuclear processes, but not in the same way the Sun does. Supergiant stars fuse hydrogen into helium in a shell surrounding a core which fuses other, heavier elements (like helium) into even heavier elements (like carbon, oxygen, and nitrogen). As the supergiant ages, its core gets hotter and denser allowing even heavier elements to fuse. This continues up to the fusion of chromium into iron. After the chromium fuel in the star has been used to create an iron core, the process stops. The reason for this is that the process of fusing iron or fissuring iron (breaking it apart) takes energy, rather than producing energy. Since no energy is produced by changing iron (either into to a smaller nucleus or a larger nucleus), the star becomes unbalanced again. No energy is being produced, so the force needed to balance gravity and prevent collapse is gone and a catastrophic event, called a **supernova** occurs. Supernovae will be discussed in much more detail in chapter 3.

The least luminous stars are about 100,000 times less luminous than the Sun. It may be a surprise to learn that the least luminous stars are *not* the smallest stars. Rather, these stars are very low mass stars that produce energy essentially the same way our Sun does—through the process of thermonuclear fusion of hydrogen into helium. These stars have low luminosities primarily because they have very low surface temperatures, but also because they are small (but not the smallest). The relationship between luminosity and size and surface temperature will be discussed in much more detail later in this chapter.

TEMPERATURE OF STARS

The hottest known star (zeta Puppis) has a surface temperature of almost 40,000 K. That is just over seven times the Sun's surface temperature. This kind of star is producing energy by fusing hydrogen into helium through thermonuclear fusion in its core, but the details of the process are a little different from the process in the Sun. This star is much more massive than our Sun, and, therefore, has much more hydrogen to fuse; also, it uses up this vast supply at a much faster rate than our Sun does.

The coolest known star (eta Muscae) has a surface temperature of just under 3,400 K. This star is also producing energy by fusing hydrogen into helium through thermonuclear fusion in its core, much like our Sun. This cool star, however, is much less massive than our Sun and, therefore, has much less hydrogen to fuse. Unlike the Sun, this star will fuse hydrogen with a very high efficiency resulting in a longer life than one might expect for such a small star. For the star to live so long, it will fuse almost 100 percent of the hydrogen it contains into helium, using convection processes to transfer fuel from near the surface of the star into the core for fusion. Interestingly, the range of surface temperatures of stars is much smaller than the ranges of the other properties discussed.

COLORS OF STARS

In addition to having many different possible masses, sizes, luminosities, and surface temperatures, stars also can be different colors. Our Sun appears white or yellow to us. That is because our Sun emits most of its light at an almost equal amount across all the wavelengths of the visible spectrum. You may already know that when all the colors of light are mixed, the color of the light is white.

A star's color has to do with what colors of light the star emits. If a star emits more of its light in the red part of the visible spectrum than the blue part of the visible spectrum, the star will appear red. And, if a star emits more of its light in the blue part of the visible spectrum than the red part of the visible spectrum, the star will appear blue. For the most part, stars are really only red, yellow, white, and blue. Many amateur astronomers have trained themselves to be able to distinguish these colors when viewing a star through a telescope. (Most professional astronomers don't look through telescopes at stars anymore, they rely on digital instruments to give them objective data to analyze.)

It turns out that a star's color is closely related to its surface temperature. Due to this fact, the surface temperatures of many stars were measured before astronomers even understood how stars worked or of what they were made. The relationship between color and surface temperature of stars is discussed in more detail later in this chapter.

THE STAR ZOO

Stars have been generally grouped into categories that describe their sizes. There are **giant stars** and supergiant stars (all stars with very large sizes). Generally, these are more luminous than the other stars, but their masses and surface temperatures can fall almost anywhere in the ranges described. The stars that are like our Sun, in terms of how they produce energy, are called **main sequence stars.** These stars tend to stay close to one another in size (usually not more than 10 times different from the Sun), but vary in surface temperature and luminosity widely. Finally, the **white dwarf stars** (very small in size) tend to be less luminous and at the same time, strangely hot (very high surface temperatures).

WHY DO STARS HAVE DIFFERENT COLORS?

There are stars that are red, stars that are blue, stars that are yellow, and stars that are white. How can stars have all these different colors? Why aren't all stars just yellow, like our Sun? Or, why aren't all stars just white? Does a star have the same color all the time? Are there stars that change colors?

THE ELECTROMAGNETIC SPECTRUM

Before we can really discuss why stars have different colors, we need to discuss light. Light is the same thing as electromagnetic radiation. Electromagnetic radiation is a phenomenon of nature that is caused by accelerating charged particles. This phenomenon is called electromagnetic radiation because a charged particle has an electric field associated with it (this is why a charge repels like charges and attracts opposite charges). When a charge is moving, the electric field creates a spontaneous magnetic field (this is why your compass moves when you put it near a current-carrying electric wire and why a gadget with magnetic memory, like a VHS tape or a memory stick, will be erased if left near a strong electrical or magnetic source, like a television set or a computer monitor). Because light is a result of an accelerating charged particle, and because accelerating charged particles have both electric and magnetic fields associated with them, light is called electromagnetic radiation.

Electromagnetic radiation, however, is not *just* the light we see with our eyes. The light we see with our eyes is only a small fraction of what electromagnetic radiation actually is. When astronomers talk about the full range of possible electromagnetic radiation, they call it the **"electromagnetic spectrum."** Different kinds of electromagnetic radiation are categorized by their wavelengths and/or energies. The wavelengths associated with visible light (the light we see with our eyes) range from 400 to 700 nm.

There are many other kinds of electromagnetic radiation. Some, you may have heard of before. **Infrared** is electromagnetic radiation with *longer* wavelengths and *less* energy than visible light (usually, this is considered to be the light with wavelengths from 700 nm to 1,000,000 nm or 1 mm). Infrared radiation is most often associated with the military use of night-vision goggles.

••

Night-Vision Goggles

The human body is at a temperature of about 300 K. An object with a temperature of 300 K emits electromagnetic radiation that has a peak wavelength in the infrared part of the electromagnetic spectrum. So, the reason night goggles work to see humans is because humans radiate electromagnetic radiation and the peak wavelength of that radiation is in the infrared. That is, *most* of the electromagnetic radiation that humans radiate is in the infrared part of the electromagnetic spectrum.

••

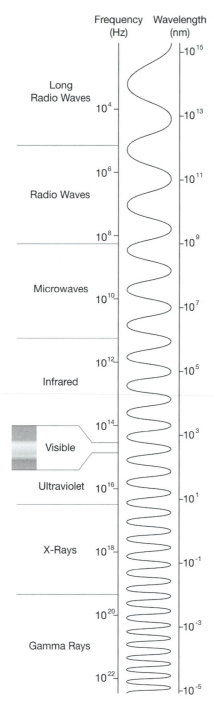

Figure 1.3 This is a diagram of the electromagnetic spectrum. There are seven main types of electromagnetic radiation; each is defined by a wavelength range. These types are called (in order from shortest to longest wavelength) gamma rays, x-rays, ultraviolet, visible, infrared, microwave, and radio radiation. Shown here are the different wavelengths that define each region of the electromagnetic spectrum. [Jeff Dixon]

Ultraviolet is electromagnetic radiation with *shorter* wavelengths and *more* energy than visible light. The ultraviolet part of the spectrum is defined to be electromagnetic radiation with wavelengths from 10 nm to 400 nm. Ultraviolet radiation is most often associated with overexposure of human skin to sunlight. The more energetic ultraviolet radiation is, the more dangerous it is to living things. Also, the more energetic ultraviolet radiation is harder to stop—it can pass through layers of clothing, for example.

....

Black Lights

Ultraviolet radiation is also the source of so-called black lights. Black lights have been popular at different times for use in personal homes and nightclubs. When black lights are used, white clothing glows. This is because the white pigment in the clothing reflects all light. Black lights don't appear to give off much light because the peak wavelength of the light emitted is in the ultraviolet part of the electromagnetic spectrum. Since the human eye cannot detect ultraviolet radiation, the human eye does not perceive the majority of the radiation coming from a black light. The reason one can see *any* light coming from a black light bulb is because the source of the light is an object that emits light at all wavelengths, even though the majority of the light emitted is in the ultraviolet part of the electromagnetic spectrum.

....

The shortest wavelength part of the electromagnetic spectrum is the **gamma rays.** A gamma ray is the most energetic form of electromagnetic radiation. The wavelength range for gamma rays is 1 fm to 0.1 nm (fm is the abbreviation for femtometer which is 10^{-15} m). The most well-known practical use for gamma rays is radiation therapy for cancer patients. This extremely energetic electromagnetic radiation is lethal to most living cells. By targeting cells that are cancerous, the hope is that repeated bombardment of the cancer cells will kill the cancer and not too many of the patient's healthy cells.

....

Photographic Film

Photographic film is a transparent film covered with a thin layer of photosensitive chemicals. Photosensitive chemicals change their properties when light is incident upon them. In the case of photographic film, the photosensitive chemicals go from being transparent to being opaque. So, unexposed photographic film is transparent. Once light is incident upon it, the film is chemically activated and becomes opaque. This is the reason that negatives are the inverse of what is seen through the lens: dark is light and light is dark.

....

The next longest wavelength part of the electromagnetic spectrum is the **x-rays.** X-rays are very energetic as well. The wavelength range for x-rays is from about 0.1 nm to 10 nm. The most common use of x-rays is the medical one. When your doctor takes and x-ray image of you, this is what happens: A part of your body (whatever your doctor wants a picture of) is bombarded with a beam of x-ray radiation. Most of the high energy electromagnetic radiation passes right through your body. Your bones, however, are dense enough to stop a significant portion of the x-rays that pass through them.

The radiologist puts a piece of film behind the targeted area of your body. The film detects the x-rays that pass through your body. Just like the film used in a regular camera, the film is transparent until it is activated by some light. So, when you see an x-ray, it is a negative image—dark where there was light and light where there was darkness. An x-ray image shows the most dense parts of your body.

On the long wavelength side of the visible part of the electromagnetic spectrum beyond infrared is **microwave radiation.** The wavelength range for the microwave regime is from 1 mm to 1 m. This is the kind of radiation your cell phone uses, as well as the kind of radiation used in microwave ovens.

Microwaves and Cell Phones

Microwave radiation is ideal for cell phone usage because these fairly short wavelengths allow them to travel far without interference. Even dense clouds will not deter microwave radiation from getting where it's going. Microwave radiation is also ideal for cooking because when microwave radiation is incident upon a dense enough object, the microwaves will be absorbed quickly, making incremental changes in the object's temperature. It is also easy to contain microwave radiation. Microwave radiation is completely reflected by most plastic. It cannot even penetrate a plastic trash bag! This also means that if you put your cell phone in a plastic box or bag, it will not be able to receive or make calls.

The last and longest wavelength range is the **radio** part of the electromagnetic spectrum. The wavelength range for radio is from about 1 m to 100,000 km. This is the wavelength range used to transmit television signals, as well as the wavelength range used for the radio in your car.

Radio Waves

Radio waves are best for transmitting information over long distances. Their low energies make them easy to modulate (so they can carry information) and cause them to reflect off clouds. This property means that a radio signal can go a long distance and use the clouds to reflect the electromagnetic radiation and keep it going back towards the surface of Earth.

One other aspect of electromagnetic radiation that must be discussed here is what is known as the wave/particle duality of light. When scientists began experimenting with light, the experiments they did showed the behavior of light to be wavelike. That is, light can be reflected, refracted, and diffracted, and light can interfere with itself. It was not until the 20th century that scientific experiments with light began to show behavior that could not be explained by a wave model. There was no other way to reconcile the experiments than to grant light the possibility of behaving like a wave *and* behaving like a particle.

So a good model for light might be a packet of energy that moves through space (at the speed of light) like a wave. This is not an easy picture to visualize, but then, light is not an easy phenomenon to describe or even define. It is important to note, however, that light can behave *both* like a particle *and* like a wave, although light itself is *neither* a particle *nor* a wave. A well-known educator once compared the wave/particle duality to the platypus. The platypus looks both like a duck and like a beaver, yet it is neither a duck nor a beaver. Similarly, light behaves both like a particle and like a wave, yet it is neither a particle nor a wave.

Since electromagnetic radiation behaves both as a particle and a wave, we can talk about both the particle and wave qualities of electromagnetic radiation. For example, we can talk about either the *wavelength* of electromagnetic radiation or the *energy* of a **photon.** A photon is a particle of light. In the discussion of different kinds of electromagnetic radiation above, the definitions are based on wavelength. Electromagnetic radiation (or light) is often classified by a wavelength, but it can also be classified by its energy. The mathematical relationship between wavelength and energy is:

$$E = \frac{h\,c}{\lambda}$$

where h is the Planck constant, c is the speed of light in m/s, E is energy in Joules, and λ is wavelength in m.

The most important part of this relationship is that energy and wavelength are related inversely. That is, for larger values of λ (wavelength), E (energy) gets smaller; and for smaller values of λ (wavelength), E (energy) gets bigger. In other words, long wavelength photons have less energy than short wavelength photons. If we put that in the context of the visible range of the electromagnetic spectrum, it means that red light has less energy than blue light.

THE IDEAL BLACKBODY

A **blackbody** is a theoretical object that absorbs all radiation incident upon it. As a result, the blackbody gets heated (gains thermal energy). The blackbody, then, because it has been heated, radiates electromagnetic radiation at every wavelength. A **blackbody spectrum** is a plot of intensity versus wavelength of the electromagnetic radiation emitted by a blackbody.

As you can see in the figure above, blackbodies with different temperatures have different and distinct blackbody curves. There are two main things that should be noted about the differences between the blackbody curves of two objects with the same size, but with different temperatures. First, the hotter blackbody has a higher peak than the cooler blackbody. Second, location of the peak of the curve of the hotter blackbody is closer to the shorter wavelength end of the plot.

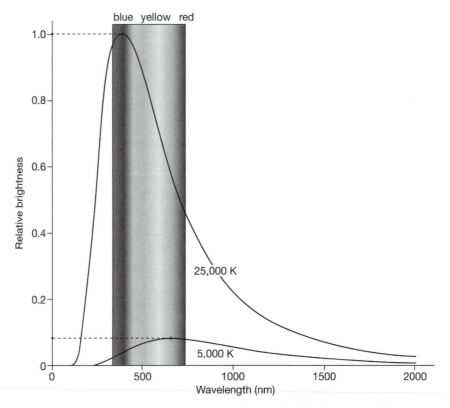

Figure 1.4 The above diagram shows two different blackbodies with different temperatures. The hotter blackbody has a curve that peaks at a shorter wavelength than the cooler blackbody. Since wavelength increases from left to right in this diagram, the curve that peaks to the left is the one that belongs to the hotter blackbody and the curve that peaks to the right is the one that belongs to the cooler blackbody. In this diagram the hotter blackbody has a temperature of 25,000 K and the cooler blackbody has a temperature of 5,000 K. [Jeff Dixon]

WIEN'S LAW

Wien's Law is simply a relationship between the temperature of a blackbody and the wavelength of the peak emission by that blackbody. The relationship is an *inverse relationship* (just like the relationship between wavelength and energy for electromagnetic radiation). This means that, just as was observed above, hotter blackbodies emit most of their electromagnetic radiation at shorter wavelengths. In mathematical language, Wien's Law is as follows:

$$\lambda = \frac{2.9 \times 10^{-6}}{T}$$

Here, λ is the wavelength at which the emission of the blackbody is at its maximum in nm, and T is the temperature of the blackbody in Kelvin. Because the mathematical relationship is inverse, as the temperature of the blackbody

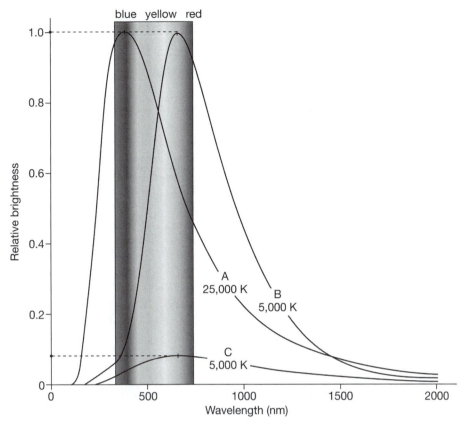

Figure 1.5 In this diagram, blackbody B (right-most peak wavelength) has a lower temperature and the same luminosity as blackbody A (left-most peak wavelength). Since this can only be true if the cooler object is larger, the blackbody curves cross one another. In cases where the curves do not cross, the blackbody sizes may be the same or very close to one another. [Jeff Dixon]

increases, the wavelength at which the electromagnetic radiation emission of the blackbody is at its maximum decreases. The opposite is also true: as the temperature of the blackbody decreases, the wavelength at which the electromagnetic radiation emission of the blackbody is at its maximum increases.

In the diagram above are three blackbody curves for three different blackbodies A, B, and C. Two of the three have the same temperature and two have the same peak intensity. The two with the same temperature have the same peak wavelength. Recall that the *peak wavelength* is the location on the horizontal axis (x-axis) that indicates where the highest point on the curve lies, whereas the *peak intensity* is the location on the vertical axis (y-axis) that indicates where the highest point in the curve lies.

The two blackbodies with the same peak intensity do not have the same temperature. These two curves rise to the same height or location on the vertical axis, but the location on the horizontal axis where this peak occurs is not the same. So, one of these two blackbodies has a lower temperature than the other. Yet, the two blackbodies are emitting the same amount of total electromagnetic radiation. Does this sound impossible? Not really. The amount of total electromagnetic radiation emitted by a blackbody is not only dependent on its *temperature,* but also on its *size.* The next section of this chapter discusses this relationship in much more detail.

THE STEFAN-BOLTZMANN LAW

The **Stefan-Boltzmann Law** describes the relationship between temperature, luminosity, and size of a blackbody. As mentioned before, luminosity is a measure of the total electromagnetic energy released by an object. In terms of a blackbody curve, we can think of luminosity as the area under the curve. So, in the diagram above the luminosity of blackbody A is greater than the luminosity of blackbody C. What else can we tell by looking at these blackbody curves?

Looking at the blackbody curves and using what we have learned about Wein's Law, we can tell which blackbody has the higher temperature. In this case, the blackbody with the higher temperature is also the one with the greater luminosity. So the hotter star emits more electromagnetic radiation. In the last example of the previous section, however, blackbodies A and B had the *same* luminosity, but *different* temperatures.

To better understand how the luminosity of a blackbody depends on its size and temperature, one should examine the mathematical relationship known as the Stefan-Boltzmann Law. The mathematical expression used to describe the Stefan-Boltzmann Law is as follows:

$$L = 4\pi R^2 \, \sigma \, T^4$$

Here, L is the luminosity in J/s, the quantity $4\pi R^2$ is equal to the surface area of the blackbody (presumably a sphere) in m^2, σ is a constant (known as the Stefan-Boltzmann constant) and T is the temperature of the blackbody in K. There are no inverse relationships here, so all quantities are *directly* related. This means that if we increase the luminosity of a blackbody, one of three things could happen: first, the size of the blackbody could increase and the temperature stays the same; second the temperature could increase while the size stays the same; and third, both the size and temperature could change in a way that causes the luminosity to increase.

It is not true that both the size and temperature must increase if the luminosity increases. The exponents in the relationship are an indication that this relationship is more complicated than the others presented in this chapter. Since the luminosity is proportional to the radius to the second power and also proportional to the temperature raised to the fourth power, we have to take these powers into consideration in our analysis.

Raising a number to the fourth power is the same as raising to the second power two times (squaring a square). So a small change in temperature will result in a much larger change in luminosity than the same change in size. That is, the luminosity is *much* more sensitive to changes in temperature than changes in size.

To make this a little more concrete, let's use some numbers. Say, blackbody A has a temperature twice that of blackbody B, but they have the same size. How will their luminosities compare? Well, to figure this out, we will look at the Stefan-Boltzmann Law for each blackbody:

$$L_A = 4\pi R_A^2 \, \sigma \, T_A^4$$

$$L_B = 4\pi R_B^2 \, \sigma \, T_B^4$$

Now, according to the information we started out with, $R_A = R_B$ and $T_A = 2T_B$. Substituting these relationships into the two equations above, we have

$$L_A = 4\pi R_B^2 \, \sigma \, (2T_B)^4$$

$$L_B = 4\pi R_B^2 \, \sigma \, T_B^4$$

Now we can relate L_A and L_B to one another. Since 2^4 is 16, we can factor that out and we have

$$L_A = 16(4\pi R_B^2 \, \sigma \, T_B^4) = 16 \, L_B$$

So if we have two blackbodies of the same size and blackbody A has a temperature twice that of blackbody B, blackbody A will be 16 times more luminous than blackbody B. What if blackbody A and blackbody B have the same temperature, but blackbody A is twice as big as blackbody B?

Now we have

$$L_A = 4\pi(2R_B)^2 \, \sigma \, T_B^4$$

$$L_B = 4\pi R_B^2 \, \sigma \, T_B^4$$

Since 2^2 is 4, we can factor out the four and rewrite these equations so that:

$$L_A = 4(4\pi R_B^2 \, \sigma \, T_B^4) = 4L_B$$

So, for two blackbodies with the same temperature, the blackbody that is two times larger will have a luminosity that is four times greater. Recall that for two blackbodies of the same size the blackbody with a temperature twice that of the other will have a luminosity that is 16 times greater. Therefore for blackbodies, while luminosity depends on both size and temperature, a change in temperature has a much greater effect than an equal change in size.

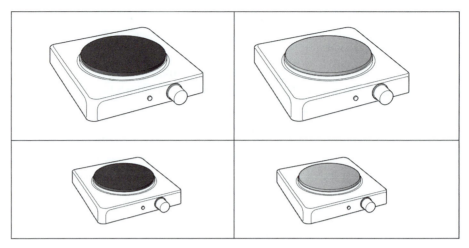

Figure 1.6 In this diagram there are four hotplates: two large and two small. One of the large hotplates is set on a high temperature (shown as black) and one is set on a low temperature (shown as light gray). The same is true for the smaller-sized hotplates. [Jeff Dixon]

According to the Stefan-Boltzmann equation, the luminosity of a blackbody depends both on the temperature and the size of the blackbody, itself. So, we can have two blackbodies with different temperatures and the cooler one could have a higher luminosity if it were sufficiently big.

The basic underlying principle behind the Stefan-Boltzmann Law is probably already a part of your intuition. Here is a more real-life example that demonstrates this claim. Consider four hotplates: two large and two small. For each size, we have one set on high and one set on low.

In the discussion below, heat is a parallel to luminosity referred to in the Stefan-Boltzmann Law. Compare a large hotplate on high with a small hotplate on high. Both hotplates have the same temperature, but which one will give off more heat? The large one will give off more heat, by the Stefan-Boltzmann Law. Now, compare a small hotplate on high to a small hotplate on low. Which will produce more heat? The one with the higher temperature, the one on high, should produce more heat according to the Stefan-Boltzmann Law. We would come to the same conclusion comparing the two large hotplates (one on high, one on low). What about a small hotplate on high compared to a large hotplate on low? Can you tell which one will give of more heat without knowing anything more specific than one is larger and one is hotter?

To determine the correct answer to the last comparison, you need more information. Whatever your instinct tells you, the Stefan-Boltzmann Law says that the dependence on temperature is more important (T is raised to the fourth power) than the dependence on size (R is raised to the second power). So, unless you know the actual temperature difference between the "high" and "low" settings on your hotplates as well as the actual size difference between the "small" and "large" hotplates, you can't tell which will give off more heat. They might even give off the same amount of heat.

STARS AS BLACKBODIES

Stars are known to physicists and astronomers as quasi-blackbodies. This means that we can assume that a star's fundamental properties (temperature, luminosity, and size) are related to one another in the same way they are for blackbodies. That is, Wien's Law and the Stefan-Boltzmann Law apply to stars, as well as to blackbodies.

To return to our question: why do stars have different colors? It is because stars are like blackbodies. The temperature of the star, therefore, by Wien's Law, *defines* the color that each star will appear to be. So, stars have different colors because stars have different surface temperatures. Since there is a range of possible surface temperatures for stars, there is a range of possible colors for stars.

The hottest stars have colors near the short wavelength end of the electromagnetic spectrum. Since our eyes only detect visible wavelength electromagnetic radiation, these stars appear blue. Stars with the lowest surface temperatures have colors at the long wavelength end of the electromagnetic spectrum. To humans, these stars appear red. So, for stars, hot is blue and cold is red—the exact opposite of what we are used to in our daily lives.

••

Wavelength and Energy

It may be interesting to point out that the relationship between wavelength and energy indicates that shorter wavelength electromagnetic radiation is more energetic. So, it would seem that stars that produce more of these shorter wavelength photons ought to have more energy, and thus, higher temperatures. It turns out they do, as shown by Wien's Law. Thinking about the relationship between wavelength and energy is a good way to make sense of Wien's Law. Hot blackbodies should have a lot of energy, so they should be able to produce and should preferentially emit high-energy photons, so the wavelength at which a hot blackbody emits most of its electromagnetic radiation should be very short. According to Wien's Law, it is!

••

As we will discuss in much more detail in chapter 3, the surface temperatures of stars change throughout their lifetimes. A star will decrease its surface temperature as it changes fuel sources going from a bluer color to a redder color. Later, a star may increase its surface temperature going from a redder color to a bluer color. So, stars do not have the same temperature or color for their entire lifetimes. That is, a star will change color during its lifetime getting first redder, then bluer in color.

In the beginning of this section it was mentioned that stars can be red, yellow, blue, and white. Why not purple, orange, or green? The colors in the visible part of the electromagnetic spectrum are red, orange, yellow, green, blue, indigo, and violet. Technically, a star could be any of these colors or any color in between. However, our eyes are not able to detect all of these colors in light from distant stars. And, even though the peak wavelength of a star may be green, if the peak is essentially flat within the visible part of the electromagnetic spectrum, we will see white because all colors will be equally bright. In fact, most of the stars we see appear to be white because, even though the peak wavelength of the electromagnetic radiation coming from the star is somewhere within the visible part of the electromagnetic spectrum, the peak is essentially flat within the visible part of the electromagnetic spectrum.

On the other hand, when a blackbody curve is distinctly lopsided within the electromagnetic spectrum, there is a clear differentiation of color. A star

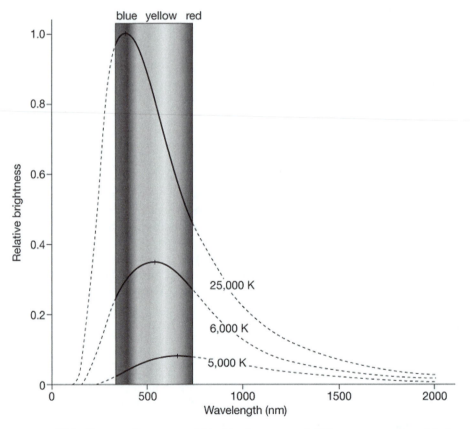

Figure 1.7 This diagram shows several blackbody curves of different-temperature blackbodies, highlighting the shape of the curve as it passes through the visible part of the electromagnetic spectrum. This aspect of the blackbody curve affects how humans perceive the color of the blackbody. A blackbody that has a curve that is higher on the red end of the visible spectrum appears red. A blackbody that has a curve that is higher on the blue end of the visible spectrum appears blue. A blackbody that has a curve that is flat across the visible spectrum appears white. [Jeff Dixon]

will appear red or blue. (Actually most stars that are called "red" look quite orange.) So, while in theory a star can have any color, in practice there are really only four distinct colors observed in stars: red, blue, yellow, and white.

SPECTRAL CLASSIFICATION

Color was the easiest way to group stars, at first. Late in the 19th century, however, telescopes were equipped with spectrometers. These instruments allowed astronomers to separate the light coming from an astronomical object into its many different wavelengths. Immediately, astronomers knew there was more to learn about stars. In the late 19th century, it was thought that stars were perfect blackbodies. The spectrum of a perfect blackbody should look like a solid rainbow; however, when astronomers looked at the spectra of stars, there were dark lines in the spectra. The spectra they observed were not at all what one would expect for a perfect blackbody; however, the spectra of other stars had some similarities to the spectrum of the Sun.

Absorption spectra caused other scientists to attempt to explain how the dark lines could be formed, given what was understood about stars and the Sun and how such spectra are formed. A culmination of many years of work by many different and unexpected people (not all professional astronomers!) unraveled the mystery of the dark lines in the spectra of stars and revealed many new things about what stars are and how they work.

SPECTRA

To start out this discussion and to tell this amazing story, it is essential that the notion of an astronomical spectrum is well understood. A spectrum (the plural of spectrum is spectra) is an image of a light source taken through a prism or some other light-dispersing medium. This kind of image is also called a "spectrogram," in astronomy. A spectrum shows the light separated into its different wavelengths, like a rainbow.

The first scientist to attempt to understand the dark lines in a star's spectrum was Joseph von Fraunhofer in 1814. Fraunhofer observed the dark lines in the Sun's spectrum through a prism. He made careful measurements of the wavelengths of these lines and even designated them with letters. Fraunhofer did not understand why the lines were present in the Sun's spectrum; however, he determined that these lines were not an artifact of his prism, but, rather, a natural part of the Sun's spectrum.

....

Joseph von Fraunhofer

Joseph von Fraunhofer was born to a glassmaker (glazier) in 1787. By the time he was 11 years old, both his parents had died and he had become an orphan. He was taken on as an apprentice to another glazier, Philipp Weichelsberger. Through an unusual circumstance, only a few years later, Fraunhofer met and was taken under the wing of the Prince Elector of Bavaria. The prince required that Weichelsberger allow Fraunhofer time to study. Within a short period of time following this, Fraunhofer went to work at the Optical Institute in Benectkbeuern (a monastery devoted to glass-making). His study of the Sun began in 1814 with the invention of his own spectroscope. In 1818, after having discovered how to make the best optical glass in the world and having invented very precise methods for measuring the dispersion of light, Fraunhofer became the director of the Optical Institute. In 1821, Fraunhofer invented the diffraction grating, which allowed him to control the dispersion of light and measure very accurately the wavelengths of the dark lines he observed in the Sun's spectrum.

In the 19th century, Robert Bunson and Gustav Kirchoff invented the spectroscope. This instrument was used to observe light from various sources. With this instrument, scientists learned about emission and continuous spectra. Bunson and Kirchoff discovered that every chemical element had an associated set of dark or bright lines that identified the element uniquely. They deduced, correctly, that the absorption lines seen in the Sun's spectrum were caused by the presence of certain elements in the atmosphere of the Sun and the atmosphere of Earth.

In the early 20th century, the spectroscope was adapted to disperse the star light received by telescopes, adding a photographic plate to record the spectra observed. These instruments are called spectrometers. It is with the astronomical tools of the 20th century that the spectra of hundreds of thousands of stars were observed and studied, bringing to astronomy a deeper understanding of the nature of stars.

Spectrometers

A spectrometer is a system of lenses and other optical elements (a grating or prism and mirrors) that separates the light coming from a star. The light from a star first passes through a slit; this creates a narrow beam of light to be dispersed by the next element. Next, the narrow beam of light passes through a prism or diffraction grating. Both of these types of elements separate the light by wavelength.

The prism refracts the light in the narrow beam. The light is bent a little as it passes through the prism because the speed of light inside the prism is different from the speed of light in air. (The speed of light changes in every medium; in general, the denser the medium is, the slower the speed of light.) The index of refraction of a medium is actually the ratio of the speed of light in a vacuum divided by the speed of light in the medium in question. For this reason the index of refraction of a medium is usually a number greater than one. The index of refraction of the prism is greater than the index of refraction of air.

Even though all wavelengths of light travel at the same speed within a medium, one property of refraction is that different wavelengths of light are

Measuring the Speed of Light

The speed of light was first measured in 1676, by Ole Römer, a Danish astronomer. He measured the speed of light using Io, one of the moons of Jupiter visible from Earth with a small telescope. He used the fact that the distance between Jupiter and Earth varied with time (due to their respective orbits around the Sun). When Earth and Jupiter were furthest apart, the delay in the time an eclipse of Io by Jupiter occurred was the greatest. Although this was not a direct measurement of speed, if the distance between Jupiter and Earth is known, the time delay can be converted to a speed. (Speed, by definition, is distance traveled divided by elapsed time.)

In the 17th century, the distances to the planets and the Sun were not known, but relative distances had been determined through geometrical analysis and observations. In the 1670s, a measurement of the parallax of Mars was attempted. The parallax of Mars (the difference in position of Mars relative to background stars when viewed simultaneously from different positions on Earth) was used to determine the distance from Earth to Mars and then from Earth to the Sun. Knowing the distance between Earth and the Sun is the key to knowing all the distances in the solar system, since they were known in astronomical units. An astronomical unit, by definition, is the distance between Earth and the Sun.

Römer measured the delay in the time of an eclipse of Io by Jupiter when the distance between Earth and Jupiter was the greatest. Using the early measurements of the astronomical unit to determine the actual distance between Earth and Jupiter, Römer calculated the speed of light to be 125,000 miles per second.

About 50 years later, in 1728, the English astronomer James Bradley attempted to measure the speed of light. He used the motion of the Earth orbiting the Sun to determine the speed of light. The angle at which the light from a star will appear to vary due to the motion of the Earth is dependent on the speed of light. Think of the angle as an angle in a right triangle. The sides of the triangle are distances proportional to the speed of the Earth and the speed of light, respectively. By this time, telescopes were accurate enough to measure such small angles. Bradley was able to measure the speed of light to be 185,000 miles per second, much closer to its actual value of 186,000 miles per second.

Refraction

Refraction is a property of waves. Waves are bent when they pass through a medium in which their speed changes. So, light, behaving as a wave, travels slower in glass, for example, than in air. Therefore, light passing through water is refracted. This is why a solid stick looks broken when viewed above and below the water simultaneously.

bent different amounts. This is because what changes is not exactly the speed of the photon, but the phase velocity of the wave front, causing the light to be spread out by wavelength, like a rainbow. This effect is what makes a spectrograph "work"—this is what causes the spectrum.

The diffraction grating accomplishes the same end, but by a different means. **Diffraction** is the bending of waves through **interference** of the waves with one another. Interference causes some waves to cancel out with one another while others add up. This creates a pattern of light and dark regions for each wavelength. The bright regions for consecutive wavelengths

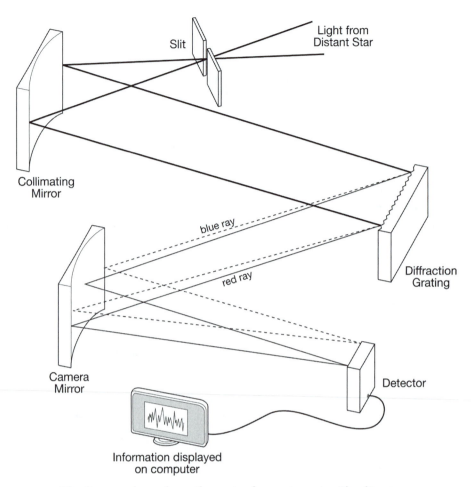

Figure 1.8 The diagram above shows the parts of a spectrometer. The slit acts as an aperture for the light, limiting the source to the light that can enter the slit directly. The diffraction grating or prism spreads out the light and separates the light by wavelength. The collimator lines up the rays of light so they are parallel. Finally, the detector measures the number of photons at each point in the spectrum. [Jeff Dixon]

appear adjacent to one another making an image like a rainbow. So, in other words, the diffraction grating separates the light by its wavelengths using the interaction of light with itself. This effect is what causes the spectrum and what makes the spectrometer work.

Diffraction and Interference

Diffraction and interference are also properties of waves. Interference occurs when two waves are superposed one upon another. When this happens the waves interfere constructively, giving the wave higher amplitude (making the light brighter), or destructively, giving the wave zero amplitude (making the light disappear). Diffraction occurs when a wave passes through a very narrow gap.

A diffraction grating is an optical element that allows light to pass through many (usually several thousand or more) tiny (smaller than the wavelength of the light being transmitted) gaps. When light waves pass through a diff raction grating the waves are bent and spread. In addition the light waves pass-ing through a diff raction grating interfere with one another. Th e combination of all these processes occurring is called diff raction.

Three Kinds of Spectra

Different physical processes can be identified simply by observing the spectrum of an object with a spectrometer. There are three distinct kinds of spectra that define three distinct physical processes. The three kinds of spectra are called continuous, emission, and absorption spectra.

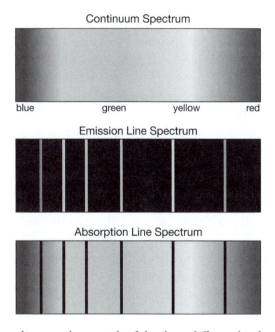

Figure 1.9 The above diagram shows each of the three different kinds of spectra. A continuous spectrum looks just like a rainbow of uninterrupted colors. This type of spectrum is formed by a blackbody source, like a hot solid or a hot, dense gas. An absorption spectrum is a continuous spectrum, but with dark lines (absorption lines) where a small range of specific wavelengths is missing. This type of spectrum is formed when the light from a blackbody passes through a low density, lower temperature gas. An emission spectrum is the opposite of an absorption spectrum; most of the length of the spectrum is dark (no light), only lines of different colors are visible, and the number and colors of the lines depend on the chemical composition of the emitting source. This type of spectrum is formed in a hot, low-density gas. A blackbody may be the source of the heat, but the gas must be hot enough that the atoms that make it up are mostly ionized or very excited. In most cases in astronomy, this type of spectrum is indicative of matter in its plasma state (ionized gas). [Jeff Dixon]

A continuous spectrum looks like a rainbow. It is a continuum of colors where every wavelength of light is represented. This kind of spectrum is what one would see if one were looking at a perfect blackbody. The light from the filament of a lightbulb has this type of spectrum.

An **absorption spectrum** looks like the continuous spectrum with dark lines in it, so that some wavelengths appear to be absent. Sometimes this is called a "dark line" spectrum. An absorption spectrum can be seen when a low temperature cloud of gas is in between an observer and a blackbody. In this case, some of the light from the blackbody that was traveling towards the observer gets absorbed by the intervening gas. But, since the gas is cool compared to the blackbody (so that most of the photons coming from the blackbody are energetic enough to excite the gas, but not **ionize** it), all the electrons in the gas are bound to their nuclei. Because of this, the gas can only absorb certain specific wavelengths of the light. The wavelengths absorbed by the cloud correspond to the energy differences between the energy levels in the atoms that make up the gas cloud.

Typically, in astronomy, the **Bohr model** of the atom is used as a reference to understand how light effects gas, creating different kinds of spectra. The Bohr model describes an atom as a nucleus with electrons in orbit around the nucleus. The orbits in the Bohr model are circular. This is where the model is inaccurate—actually, the electrons move in quasi-chaotic orbits around the nucleus and the orbits have several different shapes that define them. For the purpose of understanding the formation of spectra, though, the Bohr model is complete enough. The shape of the orbits is irrelevant to this discussion. Rather, what is important is that each orbit represents a different energy state for the electron. In this aspect, the Bohr model is consistent with modern atomic theory.

Each energy level in the atom, as defined in the Bohr model, describes an "allowed" state for the electrons in that atom. In the case of hydrogen, there is only one electron, but the energy levels still define energy states in which that electron can exist. In a hydrogen atom, the electron *cannot* have an energy that is in between any two of the energy levels. So, any process that would increase the energy of a bound electron by an amount less than that necessary to reach the next allowed state would not occur.

Each element has its own system of energy levels specific to that element. Therefore, each element has its own individual "fingerprint" of allowed energy levels. In an absorption spectrum, the dark lines are at wavelengths that correspond exactly to the energy differences between allowed energy levels in the atoms of the intervening cloud.

The third kind of spectrum is an **emission spectrum.** An emission spectrum looks like several distinct bright lines of different colors. Sometimes this is called a "bright line" spectrum. An emission spectrum is seen when a cloud of gas is heated so that the atoms in the cloud are ionized. When the atoms are ionized, each atom is missing at least one electron. So there are also many electrons that are not bound to an atom in the cloud. These electrons

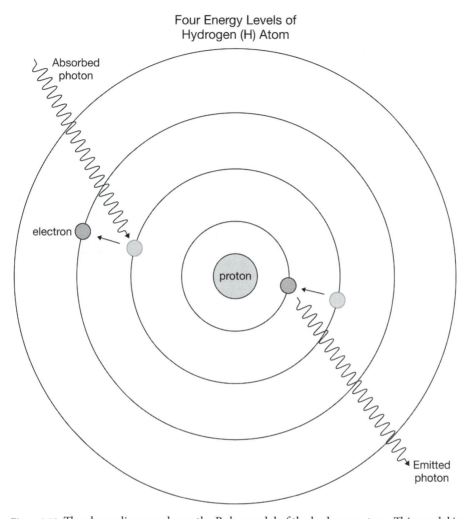

Four Energy Levels of
Hydrogen (H) Atom

Absorbed
photon

electron

proton

Emitted
photon

Figure 1.10 The above diagram shows the Bohr model of the hydrogen atom. This model is named for its inventor, Neils Bohr. Bohr hypothesized that the electron of the hydrogen atom had specific "allowed" orbits where it could be. Bohr's model shows these "orbitals." The modern use of this less-than-accurate model is in the orbitals (now called energy levels). Electrons can only have specific energies around their nuclei (for all atoms, not just hydrogen). Bohr's model allows one to calculate the wavelength of light associated with any energy level difference. Bohr's model is also good for visualizing what is meant by energy levels, although it is not an accurate physical representation of an atom. [Jeff Dixon]

combine with the ionized atoms in the clou. When the electrons recombine with the atoms, they lose energy in the form of light. The light emitted by the electrons as they lose energy and step down the levels in the atom with which they have combined, is directly related to the differences in energy between any two allowed energy levels in the atoms of the cloud. So, if the cloud is made of hydrogen, for example, the lines observed in the emission spectrum will be at wavelengths that correspond to the energy differences between the

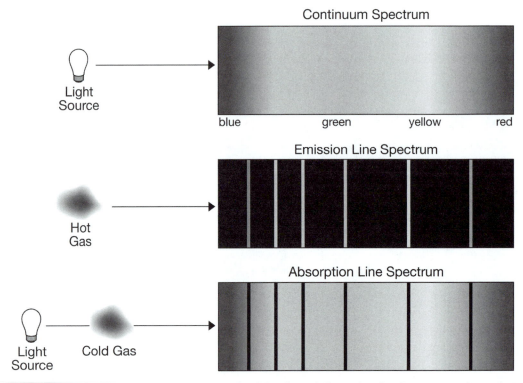

Figure 1.11 The diagram above shows how each of the three different kinds of spectra are formed. A continuous spectrum is formed by the emission of light from a blackbody (like a filament light bulb) passing through a prism or grating. An absorption spectrum occurs when light from a blackbody passes through an absorbing medium (like a cloud of gas whose atoms are not excited above the ground state) and then passes through a prism or grating. An emission spectrum occurs when an emitting medium (like a clouds of gas whose atoms are ionized or excited to a very high energy state) emits light, which is then passed through a prism or grating. [Jeff Dixon]

Neils Bohr

Neils Bohr was born to Christian and Ellen Bohr of Copenhagen, Denmark, in 1885. Christian Bohr was a professor of physiology at the University of Copenhagen, so Neils's early interest in science was nurtured and encouraged. At the age of 26, Neils Bohr received his doctorate. Following that, Bohr went to England to work with J. J. Thomson, but ended up in the lab of Ernest Rutherford. Among the many things that Bohr is known for, one lesser-known accomplishment is that he was the first to recognize was the importance of the atomic number. Bohr's atomic model stemmed from his idea that an atom could exist only in a set of discrete stable energy states. Bohr's model was able to predict the frequencies of the series of emission lines observed in the spectrum of ionized atomic hydrogen. This model was the first step towards a quantum theory of the atom. In Bohr's atomic model, the frequency of the radiation emitted is proportional to the energy difference between the two energy levels through which the electron has moved.

During World War II, Bohr and his immediate family were transported to Sweden, then to England. There, Bohr worked with the British and American scientists who eventually developed the atomic fission bomb. As part of the Atomic Energy Project, Bohr was intimately involved in creating the first bomb to exact massive human destruction. As a result, perhaps, he devoted much of his work in his later years to peace and the peaceful applications of atomic physics. In an open letter to the United Nations, he promoted openness and sharing of information as well as prevention of the use of weapons of mass destruction.

••

allowed energy levels within the hydrogen atom. The number, brightness, and colors of the lines in an emission spectrum depend on the atoms and molecules out of which the cloud is made.

It follows from their descriptions that emission and absorption spectra are the inverses of one another. That is, the lines missing in an absorption spectrum are the lines that would appear if the same cloud were hot enough to produce an emission spectrum. The lines in both the emission and absorption spectra tell us about the chemical composition of a cloud. The lines do not tell us about the heating source (if it is an emission spectrum) or the continuous source (if it is an absorption spectrum). The lines only tell us about the intervening cloud.

STELLAR SPECTRAL CLASSIFICATION

The spectrum of a star is, in fact, a combination of all three types of spectra discussed above. However, for the most part, the emission lines of most stellar spectra were not distinguishable in the first generation of stellar spectrograms. Instead, astronomers most often describe stellar spectra (even today) as absorption spectra.

In the late 19th century, the director of the Harvard Observatory at the time, Professor Pickering, realized and capitalized on the enormous potential in the educated women then graduating from women's colleges in the United States. Most of the women who chose to pursue astronomy were unemployed and looking for work. These women were well-trained and would work for almost nothing, since they had no hope of ever being able to practice astronomy. One of the many women hired was Annie Jump Cannon.

••

Stellar Spectra

Stars have absorption spectra because their atmospheres contain elements that absorb the blackbody source of light coming from the star's "surface." The atmosphere of a star is much less dense than the star's surface, so the atoms are less excited. Some atoms and molecules are even still neutral and in low energy states, depending on the star's temperature. What astronomers did not understand in their analyses of stellar spectra early on, was that the elements they observed did not make up the stars, but rather the stars' atmospheres. And, what elements appeared in the spectra had more to do with the surface temperature of the star than the composition of the star.

••

Pickering assigned Cannon the tedious job of sorting stellar spectra. The Harvard Observatory had been the recipient of the Henry Draper Memorial fund. Pickering had long ago decided to put these funds to use in his endeavor to obtain stellar spectra and classify stars by these spectra. Three women had worked on the project previous to Cannon. Two of the three developed their own classification scheme. Nettie Farrar started the project in 1886, but worked on it for only six months before quitting to get married.

The Pickering-Fleming Classification Scheme

Williamina Fleming started out working as Pickering's housemaid when her husband left her pregnant with their child. She ended up working at the observatory when Pickering became frustrated by some of his male employees who lacked the discipline and background knowledge to achieve the task he set before them. Shortly after demonstrating great potential while working as a clerk, Fleming picked up where Farrar left off. Fleming based her classification scheme on the presence of hydrogen absorption lines, or lack thereof. The stellar spectra with the strongest hydrogen absorption lines were called A stars. The B stars had the next strongest hydrogen absorption lines and so on. Fleming continued here work until 1890, when she moved on to manage the "computers" (other young women hired to perform mathematical computations) and eventually became an honorary fellow in astronomy at Wellesley College in 1907.

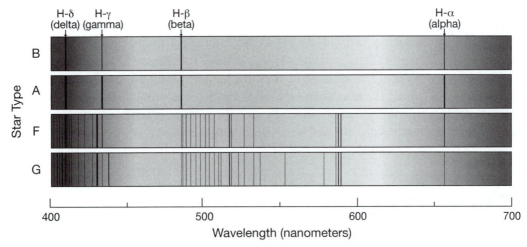

Figure 1.12 The figure above shows spectra of typical B, A, F, and G stars. In each spectrum shown, the hydrogen absorption lines are labeled. These lines are strongest in the A stars, although they are not the hottest. Other absorption lines begin to appear in the F and G type stars. This is because the atmospheres of these stars are cooler, so heavier elements are present in lower energy states. At higher temperatures, these elements are quicker to lose their outer electrons, which are the ones that would cause visible wavelength absorption lines. [Jeff Dixon]

Maury's Classification Scheme

Antonia Maury was the niece of Henry Draper. Upon graduating from Vassar College in 1887, she was hired by Pickering at her father's request. Maury developed a classification system that was based on the strength and clarity of the absorption lines found in stellar spectra. She also divided the stars into groups based on which lines were visible. To this end, she developed a method of line identification that drew mainly on what was then known about the Sun's spectrum, in addition to the spectra of several very bright stars (in Orion, for example).

Maury's classification scheme was complicated and relied on very good spectrophotometry as well as a theoretical understanding for the cause of the lines, which was not readily accepted or available at that time. (In fact, it is not clear from the publication of the classification scheme that she understood the underlying theory for why certain lines were present, while others were not; however, her scheme was the first to separate main sequence stars from giants and dwarfs based on the clarity and thickness of certain absorption lines.)

Despite these facts, Maury published her classification scheme in a catalogue containing a fraction of the stars she classified, called "Spectra of Bright Stars Photographed with the 11-inch Draper Telescope" in the Annals of the Astronomical Observatory of Harvard College. Pickering wrote a forward to the 134-page publication stating that Maury was responsible for the classifications in their entirety. Apparently, this was Pickering's politically correct way of stating publicly that he disagreed with Maury's classification scheme. Maury left the Harvard College Observatory near the time of the publication of her work, which occurred in 1896.

Annie Cannon's Classification Scheme

When Cannon took over the job of classifying stellar spectra in 1911, she developed a new classification system that was not unlike either Maury's or Fleming's classification scheme. Cannon used Fleming's letter groupings, but regrouped them and reordered them based on what Maury had done. Using Maury's scheme, Cannon condensed Fleming's 22 categories to about 10 distinct categories, and, in the process, Cannon also reordered Fleming's alphabetical categories. In the end, except for a few stars that turned out to be spectroscopic binaries or other unusual phenomena, the spectral classification scheme became O B A F G K M N R S. Further, Cannon developed a way of identifying transitional spectra by adding a digit following each letter (0–9) indicating which spectral type the star's spectra was most like.

It is Cannon's classification scheme that has survived and is used by modern astronomers. In her lifetime, Annie Jump Cannon classified over half a million stars. To this day, she is the only person to ever achieve such a feat.

Even though many have tried, there is not yet computer software that can replicate the efficiency of Cannon and her team of human computers.

WHAT ARE STARS MADE OF?

Although it was apparent that stars had different elements visible in their spectra, it was still not clear to astronomers whether the stars' spectra were different because the stars contained different elements, or whether stars were made of the same elements, but had different spectra for some other reason.

In the early 20th century, the accepted theory for planet and star formation was that planets formed from material from the surface of the Sun. For this reason, astronomers thought that stars must be composed of iron. At that time, Earth was known to contain an iron and nickel core, so it seemed a natural conclusion to draw that, given that the planets were formed from the material on the surface of the Sun, the Sun must also contain iron (and other heavy elements). But the spectrum of the Sun did not include evidence of all the elements found on Earth, or even the same distribution of elements.

MEGHNAD SAHA

Meghnad Saha enters this story with the needed solution. He read the publications from the Harvard Observatory with a critical eye and tried to explain what Maury and Cannon could not. Saha was born in 1893 in what was then India, near the city of Dhaka (now the capital of Bangladesh). Saha was recognized early in his life for his genius. He went to a village elementary school, then a city middle school, then the Dhaka Collegiate school. He ranked second in the nation on the entrance exam for college and was admitted to the Presidency College for studies in mathematics.

After accomplishing his masters of science degree in 1915 from Calcutta College, Saha was asked to teach in the newly established Science College. Although Saha wanted to study applied mathematics and physics, there was no structure within the higher education system of his country to facilitate that, so he continued his studies independently. His first major publication in astrophysics was in the *Astrophysical Journal* in 1919. This publication was his paper called "On Radiation Pressure and The Quantum Theory." His most famous scientific work "Thermal Ionization of Gases" was published in 1920. It is these two scientific papers that bring him into the story of stellar spectra.

Saha found a solid theoretical basis for the spectral sequence developed by Cannon. Saha's 1919 dissertation on radiation pressure demonstrated how the surface temperature of a star and the composition of a star's atmosphere would affect the absorption spectrum observed. A paper published in 1921, entitled "On a Physical Theory of Stellar Spectra," explained this in the context of Cannon's classification scheme. Connecting all the work done by Fleming, Maury, and Cannon, Saha was able to make sense of the new spectral classification scheme developed by Cannon.

CECILIA PAYNE-GAPOSCHKIN

In the meantime, in 1923, Cecilia Payne-Gaposchkin, a formidable woman, became the first person to receive a Ph.D. for work done at the Harvard Observatory. Cecilia Payne wrote her dissertation on "Stellar Atmospheres, A Contribution to the Observational Study of High Temperature in the Reversing Layers of Stars." Her dissertation received great acclaim. Henry Norris Russell, a famous astronomer at the time and the co-inventor of the H-R Diagram, wrote, "It is the best doctoral thesis I ever read." (Russell 1925).

To set the scene, despite the work of Fleming, Maury, and Cannon, a clear understanding for *why* stellar spectra varied was not yet part of the accepted paradigm. There were many problems. First, stellar surface temperatures appeared to be too hot to allow for absorption spectra to exist. Second, stellar spectra appeared to indicate that stars were not all made of the same material. Some stars appeared to contain no hydrogen at all, since their spectra showed no hydrogen absorption lines. At the same time, other stars appeared to be made of only hydrogen, and no other elements.

It was Payne's Ph.D. dissertation that demonstrated the homogeneity of the chemical composition of the stars. She showed, using the quantum mechanical understanding of atomic structure, that the different stellar spectra observed were due to different physical conditions, rather than a different chemical composition. Payne held back, though. Her data showed that stars were mostly made of hydrogen and helium with a very small amount of other, heavier elements. However, the paradigm of the time was that the planets formed from the outer layers of the Sun. Hence, it was thought that Sun, and other stars, therefore, should be made mostly of iron, as was Earth. Given this paradigm, Payne omitted the abundances of hydrogen, helium, and oxygen from a table describing the relative abundances of elements in stellar atmospheres.

PUTTING IT ALL TOGETHER

In the end, then, it was the work of four women at the Harvard Observatory and one man from Bangladesh who solved the mystery of the dark lines seen in the spectra of stars and the Sun. Three of the women were able to see the patterns in the strange spectra and recognized a systematic change. Documenting these changes was the first fundamental step toward understanding their meaning. The man from Bangladesh, along with the fourth woman, put the puzzle pieces together to reveal the underlying physics that made all the work of the three women who classified the spectra come to fruition.

Now, we understand that, although the spectra of stars have subtle differences, all stars are made of, basically, the same stuff: hydrogen and helium, with traces of other elements. Also, we know that the subtle changes in the spectra of the stars are due to differences in surface temperature, not differences in chemical composition. We can actually determine the surface temperature of a star, using the absorption line pattern found in its spectrum. All this leads to even more information to process about stars. If they're all made of, essentially, the same things, why are they different temperatures? Why do they have different colors? Why do they have different luminosities? These, and other questions will be addressed in the following chapters.

RECOMMENDED READINGS

Asimov, Isaac. *Asimov on Astronomy.* New York: Bonanza Books, 1988.

Hawking, Stephen. *On the Shoulders of Giants.* Philadelphia: Running Press, 2002.

Jastrow, Robert. *Red Giants and White Dwarfs.* New York: W.W. Norton and Company, 1990.

Russell, Henry Norris. Letter to Cecilia Payne-Gaposchkin's advisor Harlowe Shapley. 1925. http://www.harvardsquarelibrary.org/unitarians/payne2.html.

Seeds, Michael A. *Astronomy: The Solar System and Beyond.* 5th ed. Pacific Grove, CA: Brooks Cole, 2006.

WEB SITES

http://www.harvardsquarelibrary.org/unitarians/payne2.html

http://www.sdsc.edu/ScienceWomen/cannon.html

http://en.wikipedia.org/wiki/Antonia_Maury

http://banglapedia.search.com.bd/HT/S_0022.htm

http://www.columbia.edu/~ah297/un-esa/ws1999-letter-bohr.html

http://nobelprize.org/nobel_prizes/physics/laureates/1922/bohr-bio.html

http://en.wikipedia.org/wiki/Joseph_von_Fraunhofer

http://www.calcuttaweb.com/people/msaha.shtml

http://eo.nso.edu/MrSunspot/records

http://www.astro.wisc.edu/~dolan/constellations/extra/nearest.html

2

Star Light, Star Bright

Star light, star bright,
first star I see tonight,
I wish I may, I wish I might,
have the wish I wish tonight.

As many people already know, the first star one sees (near the western horizon as the Sun sets) is actually usually a planet (Venus or Mercury, in fact). However, this nursery rhyme actually has something to do with the way the astronomical magnitude scale (used to describe the brightness of stars) was developed. The magnitude scale is used by professional and amateur astronomers alike. Knowing a star's magnitude tells you how long to expose a digital imager to take an image of the star, or how dark the sky has to be to see the star if you want to observe the star with the naked eye.

Since astronomy has been taught, students of astronomy have been perplexed about this unusual scale. For one thing, the scale is backwards, which makes it counterintuitive. Secondly, the scale is logarithmic. Although most of the human senses actually operate on a logarithmic scale, most humans are not very comfortable with the mathematics of a logarithmic scale. The first section of this chapter will be devoted to clarifying the complications (mathematical and logical) of the astronomical stellar magnitude scale.

MAGNITUDES

Before the time when astronomers had the tools of modern astronomy that allow them to measure brightness in units of energy per second, ancient astronomers developed a way of classifying stars by their apparent brightnesses. The classification system is now called the stellar magnitude scale.

This method was apparently developed by watching the sky as the Sun set and for hours afterwards. The first stars that appeared were dubbed first magnitude stars. The second set of stars were called second magnitude stars, and so on. The faintest stars the ancient astronomers who developed this scale could distinguish were the sixth magnitude stars. Today, in most modern residential areas, the human eye can detect stars as faint as fifth magnitude. In many urban areas, one is lucky to see the brightest objects in the night sky—the Moon and the brightest planets.

At the time the magnitude scale was developed, humans on Earth could see many more stars than we can see with modern light pollution. Many of the constellations and asterisms we see in the northern hemisphere were named by these ancient people. To be sure, more stars in the sky were visible to the people of that time. The figures they saw had far more detail than we can detect today. However, the brightest stars in the sky still sketch out a rough outline of the figures described in the ancient literature.

Light Pollution

Light pollution is becoming a big problem for modern astronomers. When humans lit their homes with candles or even oil lamps, the night sky was very dark. It was easy to see the **Milky Way** in the northern hemisphere. Nowadays, one has to travel far from urban centers to view the Milky Way. With the expansion of cities over the last several decades, getting far from urban light pollution is becoming more and more difficult. Many people have never seen the Milky Way or the Big Dipper (the most well-known asterism in the northern hemisphere). Most children who grow up with urban light pollution are not even aware that there are more than a handful of stars to see in the night sky.

In an effort to light up the night, humans have taken away their ability to view the universe. This has contributed (along with the invention of the internet and cellular phone technology) to the feeling that Earth is a small, isolated world. It is only the recent explorations of our solar system by robots, images from our space-based telescopes, and brief trips of astronauts to the International Space Station that interrupt our feeling of isolation in this universe. This is, in no small part, due to the fact that humans no longer see the universe in our night sky.

To show how serious the damage is, satellite pictures of the night side of Earth have been taken. In these images, the major urban centers are apparent in every country. The United States of America is defined by the eastern seaboard (which is lit from the Florida Keys to the tip of Maine), the northern border with Canada (lit from Prince Edward Island to Vancouver), the west coast (lit from the northern border of Washington to the southern border of California), and the southern border (lit from San Diego, across the Mexican border, along the coasts of every southern state to the Florida Keys). One cannot miss a single urban center within the United States. Every one is identifiable by a bright, irregularly shaped blob on the map.

For professional astronomers this means that it is getting more and more difficult to find "dark sites" on the planet. Observatories housing world-class telescopes built on mountains in the southwestern part of the United States are already suffering multiple effects of proximity to urban centers. First, the pollutants in the air degrade the mirrors and their special coatings quickly, requiring them to be recoated and repolished frequently. Second, the light pollution means that visibility at these sites is reduced, requiring larger telescopes or more remote sites.

The first major observatory in the southwest was Mt. Wilson, located in the mountains just outside of Pasadena, California. This is where Edwin Hubble made many landmark observations that changed the way astronomers understood the universe. This observatory was built in the low mountains surrounding Los Angeles, west of Pasadena. Mt. Wilson is not used for astronomical research any longer. It is surrounded by cell phone and radio towers and the light from the City of Angels has made it impossible to view anything astronomically significant for many years. In fact, Hubble's greatest discovery at Mt. Wilson was made during World War II, when the city's lights were turned off at night to protect the city from attacks from enemies from across the Pacific Ocean.

Light pollution has had such a profound effect on astronomy that many astronomers have joined an organization called the International Dark-Sky Association (IDA). This organization promotes the use of lighting that does not interfere with viewing the night sky. Some countries have adapted their recommendations in areas near astronomical telescopes to protect the science. For example, in the Canary Islands, one can find some of the darkest skies in the modern world. The island of La Palma enacted a law requiring that all lighting meet IDA standards. The telescopes housed on La Roque de los Muchachos on the island of La Palma are becoming homes to some of the most important astronomical instruments on Earth.

• •

Nowadays, astronomers use two different measurements of a star's brightness. These are called **apparent magnitude** and **absolute magnitude.** These are each related to a star's **brightness** and **luminosity,** respectively. Further, modern astronomers have also refined and extended the original magnitude scale to include negative numbers, decimal numbers, and the faintest stars ever observed (26th magnitude).

Apparent Magnitude

Apparent magnitude is a measure of how bright a star appears. This is what is referred to as a star's "brightness." Sometimes apparent magnitude is referred to as how bright a star appears to the human eye. Even though the magnitude system was defined using only the human eye as a tool, modern astronomers use film, telescopes, digital imagers, and photometers to measure apparent magnitude. So, in modern times, apparent magnitude is not limited to the human eye. Nor is it limited to the location of the human eye or other instrument being on Earth. Modern astronomers have managed to launch telescopes into space and even place some (Hipparcos and Spitzer) in orbit around the Sun, rather than Earth. So, currently, these space telescopes measure a quantity called apparent magnitude, even though they are not human and do not reside on Earth.

A star's brightness depends on how much energy the star is emitting, the size of the star, and the distance the star is from Earth. A star can appear bright because it is giving off a lot of energy, or because it is large, or because it is close. So, if the only information about a star is its apparent magnitude, it is not possible to tell whether a bright star is bright because it gives off a lot of energy, because it is large, or because it is close.

Two stars giving off the same amount of energy that are the same size, but have different apparent magnitudes must be at different distances. Knowing apparent magnitude and two out of the three qualities of energy output, distance, or size, will limit the remaining quality. For example, if two stars have the same apparent magnitudes, give off the same amount of energy, and have the same size, they must be at the same distance from Earth. And, two stars with the same apparent magnitude, same size, and same distance from Earth must be giving off the same amount of energy.

Absolute Magnitude

Absolute magnitude is a measure of how much energy a star gives off, or how bright a star actually is. This is what is called "luminosity." The total amount of energy a star gives off is a measure of its luminosity. Luminosity is not dependent on size or distance of the star. So, two stars with the same absolute magnitude give off the same amount of energy, but they may be different sizes or different distances from Earth. The absolute magnitude or luminosity of a star does not provide any information about distance to the star or radius of the star.

Alternatively, the absolute magnitude of a star is the same as the apparent magnitude of the star if it were located at a distance of 10 **parsecs (pc)** from Earth. The distance, 10 pc, was chosen as the standard because the logarithm of 10 is 1. Mathematically, this makes the relationship between absolute magnitude, apparent magnitude, and distance simpler. Also, 10 pc is a reasonable distance for a star to be from Earth. There are more stars are at or beyond 10 pc from the Sun than there are stars closer than 10 pc from the Sun. (Incidentally, since the distance between Earth and the Sun is so much smaller than one parsec [it's 1/206,265th of a parsec], astronomers assume that distance from Earth is the same as distance from the Sun for all stars and other objects that reside outside the solar system.)

THE MAGNITUDE SCALE

As noted before, the magnitude scale is backwards and logarithmic. The magnitude scale is "backwards" because on the magnitude scale, a large number indicates a faint star. This is because the magnitude scale was devised by watching as the stars appear after the Sun sets and the sky gets darker.

The scale is logarithmic because, interestingly enough, it turns out that most human senses are logarithmic. We hear change in intensity logarithmically (hence the decibel scale), we see change in intensity logarithmically, we also detect odor and taste changes in intensity logarithmically. So, although our brains don't process the mathematics easily this way, apparently we are inherently wired to detect changes in intensity on a logarithmic scale.

USING MAGNITUDES

Astronomers use magnitudes to describe the apparent brightness or luminosity of a star, or other astronomical object. While the original magnitude scale went from 1 to 6 and contained only six categories and no finer distinction than a whole magnitude, modern astronomers have extended the scale to include negative numbers and decimal numbers, and to go to fainter magnitudes than ever thought possible before. The modern magnitude scale starts with the brightest object, the Sun, and extends to the faintest object ever observed with the most powerful space telescope.

The apparent magnitude of the Sun is -26.8. The absolute magnitude of the Sun is 4.8. Compared to the stars we see in the night sky, the Sun is, by far, the *brightest* star. This is due to the fact that the Sun is so close to Earth. However, compared to the stars we see in the night sky, the *luminosity* of the Sun is near the average. The brightest stars have an absolute magnitude around -10, and the faintest around 20.

Most humans can see stars as faint as apparent magnitude 6. Most of the stars in the Big Dipper have an apparent magnitude of 3 or lower. Conversely, most of the stars in the Little Dipper have an apparent magnitude of 5 or higher. (Most people have not seen the Little Dipper, but can easily identify the Big Dipper.) The largest telescopes on Earth (and space telescopes) can detect stars as faint as apparent magnitude 26. Besides the Sun, the brightest star is Sirius with an apparent magnitude of –1.4.

An apparent magnitude difference of 1 is equivalent to a brightness factor of 2.5. An absolute magnitude difference of 1 is also equivalent to a luminosity factor of 2.5. Mathematically, this seems strange. We subtract two magnitudes to get a magnitude difference, but to translate this to a brightness (or luminosity) factor we have to divide.

Let's try a simple example: Star A has an apparent magnitude of 4 and star B has an apparent magnitude of 5. Which one appears brighter? How much brighter? To answer these questions, we have to remember many different steps. First remember that the magnitude scale is backwards. So, bigger numbers mean the star is fainter. So, star B is fainter than star A because 5 is a bigger number than 4. Next, to answer how much brighter, we must calculate the magnitude difference. We have to subtract 4 from 5. So, the magnitude difference is 1 because star B has an apparent magnitude that is larger than star A by 1 (five is one more than four). Now, a magnitude difference of 1 translates to a brightness factor of 2.5. This means that since star A is 1 magnitude brighter than star B, star A is 2.5 times as bright as star B.

The problem gets more complex when the magnitude difference is greater than one. Now, for each magnitude difference, we must multiply by 2.5. So, if we have a magnitude difference of 2, the brightness factor will be 2.5 x 2.5 = 6.25. If the magnitude difference is 3, the brightness factor will be 2.5 x 2.5 x 2.5 = 15.625 and so on. Astronomers actually use 2.512, instead of 2.5, but the approximation of the factor of 2.5 is sufficient most of the time. A five magnitude difference, by definition, is a brightness factor of 100 (this is why 2.512 is more accurate: check it on a calculator).

The use of magnitudes reveals even more information about a star when we include absolute magnitudes and compare them to apparent magnitudes. After discussing how distance effects luminosity and brightness, we will return to this complex problem.

THE INVERSE SQUARE LAW

What is commonly called "the inverse square law" is just another strange phenomenon. This one is trickier than the magnitude scale, because at first, it seems to make sense and agree with our day-to-day experience, but then the inverse logic makes it difficult for our brains to process easily.

It starts out with something you already know: things that are farther away appear fainter. This makes sense. We have all experienced this at one time or another. It even fits with a basic rule that most humans learn early in their existence: closer is more. It's really the same thing: objects farther away appear fainter means that objects closer appear brighter.

Now comes the hard part. The amount that the brightness changes by is inversely proportional to the square of the distance between the observer and the object. Sounds like a lot of math mumbo jumbo, right? Well, here it is in a plain English example: Object A and object B have the same intrinsic brightness, but object B is two times further from the observer than object A. Object A will appear $2^2 = 4$ times brighter than object B.

Let's reason this one through step-by-step. Objects A and B have the same intrinsic brightness. So, if they're next to one another, one cannot tell which is brighter, because they give off the same amount of light. But object B is two times further from the observer than object A. Further means fainter, so object B is fainter. But not two times fainter, two-squared times fainter, which is four times fainter. Object B does not appear half as bright as object A, it appears 1/4 as bright. Well, that's a little unexpected. Didn't we just learn that our eyes detect light logarithmically? Why is it squared?

Imagine a point of light spreading out in all directions. At the source, the light given off is the same in all directions. Move away from the light source by one unit (meter, AU, light year, parsec, it doesn't matter which). The light will be fainter than on the surface. In every direction from the source, the amount of light seen at that distance from the source (one unit) is the same. The light from the source is spread out over that whole sphere. (Because we are dealing with surface *area*, the number is squared. The area of anything is in square units.)

If we examine a square centimeter on the surface of a star and watch what happens to the light contained in the original square centimeter as we move away from the star, that area gets larger the farther we go out. (See figure 2.1 below.) With each unit of distance from the star, the area increases by a larger amount. At one unit from the star, we have one area. At two units from the star, the area increases to four times the size it was at one unit from the star. At three units from the star, the area increases to nine times the size it was at one unit from the star. If the light from one square centimeter is contained in the first surface (at one unit from the star), then by the time we are three units from the star, the light contained in that same area is now 1/9th the brightness it was when it started.

It is the decrease in brightness over distance that is described by the inverse square law. And, it is because of the inverse square law that astronomers use absolute magnitude as well as apparent magnitude.

Recall that apparent magnitude tells how bright a star appears, while absolute magnitude tells how luminous a star is, or how bright the star *would* appear if it were at a distance of 10 pc. So, a star could have the same apparent and absolute magnitudes. Such a scenario would occur if the star were at a distance of 10 pc from Earth. If a star's absolute magnitude is a larger number than its apparent magnitude, that would mean that the star appears brighter than it would be if it were at a distance of 10 pc. Conversely, if a star's apparent magnitude is a larger number than its absolute magnitude, that would mean that the star appears fainter than it would if it were at a distance of 10 pc.

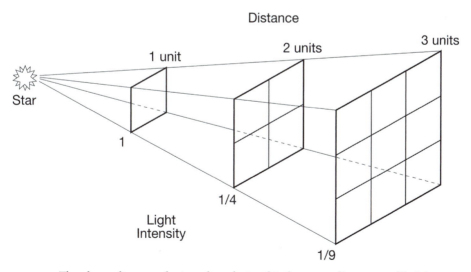

Figure 2.1 The above diagram depicts the relationship between distance and brightness. As the light travels away from the source, the surface area increases as the square of distance. So, if the light from a square travels to a distance two units from the source, the radiation will be spread out over four squares, so that each square receives one-quarter of the total light emitted by the same square at the surface. At three units from the source, the light is divided into nine squares so that each receives one-ninth of the total light emitted by a square on the surface. [Jeff Dixon]

The two different types of magnitudes are related to one another. This is, perhaps, apparent in their definitions. Mathematically, however, the relationship looks like this:

$$m - M = 5 \log d - 5$$

The lower case m is used to represent apparent magnitude, the capital M is used to represent absolute magnitude, and the d represents distance in units of parsecs.

Parsec

The distance unit parsec is a word derived from two other words: "**parallax**" and "**arcsecond.**" The parsec is defined as the distance to a star that has a parallax of 1 arcsecond. To understand, fully, what this means, we must understand both the concept of parallax and the unit arcsecond.

An arcsecond is a division of a degree of arc (there are 360 degrees in a circle). A degree is divided into 60 arcminutes. Each arcminute is divided into 60 arcseconds. So, an arcsecond is 1/3,600th of a degree.

Parallax is a method for measuring distance to stars. This method is useful for measuring the distance to only the nearest stars (out to a distance of about 100 pc, when measured from Earth). As Earth orbits the Sun, nearby stars appear to change position relative to stars that are farther away. The amount by which the nearby star's position changes is related to its distance.

On Earth, humans have experience with the parallax effect. This is why, when you watch the landscape as you sit in a moving car or train, the nearby landscape appears to move quickly, but the distant landscape appears to move more slowly or not at all. One can experience parallax simply by holding a pencil at arm's length at the center of your vision and close one eye at a time. The pencil will appear to move relative to the background in your line of sight.

To measure a star's parallax, astronomers observe the star (at least) two times so that the observations are six months apart. Since the observations are six months apart, during the second observation Earth is on the opposite side of the Sun from the first observation. This means that Earth is as far away from the position it was in at the first observation as possible. This maximizes the baseline. The angular shift of the star's position is its parallax shift. The star's parallax is equal to half its parallax shift.

To see how this relates to the star's distance, a little bit of geometry is needed. The distance from the star to Earth is the hypotenuse of the right triangle, the distance from the Sun to Earth is the short leg, and the distance from the Sun to the star is the long leg of the right triangle. The angle between the long leg and the hypotenuse is the parallax. Using some elementary trigonometry, we can easily derive the relationship that follows:

sin (parallax) = distance between Earth and Sun/distance between Earth and star

Since the parallax of even the nearest stars is in 10ths or 100ths of an arcsecond, the angle is very small, so we can use the small angle approximation that says that the sine of a very small angle is approximately equal to the size of the angle. Plugging in the distance between Earth and the Sun as 1 *astronomical unit (AU),* the equation now reads,

$$p = 1/d$$

where p is parallax (in arcseconds) and d is distance to the star. Since, by definition, a parsec is the distance to a star with a parallax of 1, the distance is in units of parsecs.

The magnitude scale is a way to quantify a star's brightness and luminosity. Using a star's brightness, one can determine its apparent magnitude. If distance to the star is also known, then absolute magnitude can also be determined.

OBSERVABLE STAR PROPERTIES

Looking at stars, astronomers can determine many of their properties or aspects. Stars have color, temperature, brightness, and size. They also have luminosity, distance, position in the sky, and motion relative to other stars. Compared with most other astronomical objects, stars are some of the easiest to learn about since so much information is observable. The only aspect of a star that is not yet observable is a star's age, which we can estimate if it is part of a cluster of stars. But to learn about a star's age, we would have to be able to observe its evolutionary sequence. As we will learn in chapter 3, the length of a star's existence is not observable because it is millions or billions of years, much longer than a human lifetime.

A lot of the aspects of a star that can be observed are properties that are not part of the star's *intrinsic* nature. For example, the distance that a star is from Earth (or the Sun) is an observable property of the star, but the distance from Earth is not a property of the star that physically defines what the star is. This section will define what the intrinsic properties of stars are and how we can use them to learn more about the physical nature of stars; it will also lay the groundwork for chapter 3, which will explain stellar evolution and describe how astronomers figured out the billion-year evolutionary sequence of stars.

INTRINSIC PROPERTIES OF STARS

An intrinsic property is one that defines an object. For example, an intrinsic property of a library would be that it contains books. An intrinsic property of a star is its luminosity, or the amount of energy the star gives off. Luminosity is not the same thing as the brightness of the star. Luminosity is the amount of energy a star gives off, whereas, brightness is the amount of energy we receive. Depending on our distance from the star, the brightness it has (or the amount of energy we receive) may be greater or smaller. Stars can appear bright because they are closer to Earth, relatively speaking. This same effect is observed on Earth when, for example, an approaching car's headlights appear brighter as the car gets closer. The car's lights are not actually giving off more light as the car gets closer; rather, the lights appear brighter because the source of the light is getting closer. (To learn more about this and the difference between apparent and absolute magnitude, see above.) The luminosity of a star, measured in absolute magnitude, is an intrinsic property of a star.

Another intrinsic property of a star is its temperature. Temperature is related to luminosity, but not simply. Luminosity is proportional to the temperature of a star raised to the fourth power. The temperature of a star is determined by examining its spectrum. The amount of energy at each given wavelength is analyzed and a peak wavelength can be determined. Once a peak wavelength is determined, Wein's Law will allow one to calculate a temperature for the star. Another way to measure temperature is by determining a stars **spectral type.** Spectral type is a classification based on the spectrum of a star that can also be used as a measure of temperature. Different temperature stars have different characteristics in their spectra. Using a classification scheme, one can identify a star's spectral type and relate that classification to a star's temperature.

In addition to temperature and luminosity, size is an intrinsic property of a star. Size is only independently measurable in certain rare circumstances (e.g., eclipsing binary star systems that are viewed exactly edge on). Such a physical scenario allows for the direct measurement of size as long as the distance to the system is known. In this case, the rate of motion of each star in the system can be determined and the size of each star can be measured by studying how the light from the system changes as each star takes a turn passing in front of the other, relative to the observer. As one star moves in front of the other, the light from the system will increase or decrease until one star is completely in front of the other. Later, when the star in front moves out of the way to reveal the star behind, the light from the system goes back to its normal level (either increasing or decreasing to reach that level). Knowing the distance to the system, the speed of the bodies can be calculated. Knowing the speed of each star, the size of each star can be calculated.

Size is also related to luminosity, as described by the Stefan-Boltzmann Law (see chapter 1), which demonstrates the relationship between luminosity, temperature, and radius of a blackbody. A star may have a lower temperature than another star with the same luminosity, but the lower temperature star must be larger than the higher temperature star to be able to produce the same amount of light.

Another intrinsic property of a star is its mass, which is a component of an object's weight and can only be independently measured in special circumstances (e.g., binary star systems). Knowing the distance to a binary star system as well as the angle of inclination of the plane of the system will allow for a mass calculation. The two stars orbit a common point in space (the center of mass of the system). Knowing the distance between the two stars (which can be derived from the distance to the system and the angle of inclination of the plane of the system) allows one to use Newton's Law of Universal Gravitation to calculate the masses of the two stars. This will be described in more detail in chapter 4. A star's mass tells astronomers a lot about how it produces energy, how long it will exist, and what it will become after it ceases to exist as a star and can no longer produce energy internally.

By unraveling the evolution of stars astronomers came to understand what properties of a star define it, that is, are intrinsic to it. Working independently, two early 20th-century astronomers Ejnar Hertzsprung (1873–1967), a Dane, and Henry Norris Russell (1877–1957), an American, accomplished this work within a decade of each other. In honor of their work, one of the most important tools used by modern astronomers for studying and understanding stars and stellar evolution is known as the Hertzsprung-Russell Diagram (commonly called the **H-R Diagram**). The Hertzsprung-Russell Diagram is pivotal in understanding what stars are physically, how they produce energy and how they evolve.

Changing Measurement of Time

In the first decades of the 20th century, communications on Earth were not nearly as speedy as they are today. In the 21st century, we measure time in milliseconds and determine winners of races to the hundredth of a second. At the beginning of the last century, two men devising the same diagram nearly 10 years apart was practically simultaneous. At the time that Ejnar Hertzsprung and Henry Norris Russell lived, communications between the thousand or so scientists who studied the universe were sent by mail, what we currently call "snail mail." And in those days, mail traveled by rail, by boat, and by the newest inventions, cars. Letters traveled around the world in almost three month's time.

THE H-R DIAGRAM

In the early 20th century, astronomers were just beginning to learn about the physical nature of the universe in which they lived. It took almost 5,000 years of written history for humans to understand that Earth was a planet in a system of planets. At the turn of the 20th century, when Hertzsprung and Russell were working, astronomers were just beginning to understand that the universe was far more vast than anyone had ever imagined.

At the Harvard Observatory, the observatory director Professor Edward Pickering had directed the women he employed to take spectra of every star in the heavens and to classify those spectra by the absorption lines visible within them (see also chapter 1). His goal was to understand what the spectra of stars could reveal to astronomers about the physical nature of stars. It turned out that he could correlate the information he gleaned from their spectra to their surface temperatures and their sizes (sometimes). With the help of other scientists who studied the physics of extremely hot and low-density gases, astronomers were able to draw the conclusion that all stars are made of the same ingredients: about 70 percent hydrogen, 30 percent helium, and a smattering of other elements.

Hertzsprung wanted to know other things. He did not have all the spectra that the women at the Harvard Observatory were studying, but he was doing spectrophotometry (measuring brightnesses of stars in different parts of the visible spectrum) on his own and used star temperature (derived from his spectrophotometry) and luminosity (in the form of absolute magnitude) to create a diagram. Russell was studying stars and measuring their distances from Earth to determine their absolute magnitudes. Russell used these data along with the Harvard women's spectra to make his own diagram.

Both diagrams showed the same unusual feature. While most stars appeared to be on a line that went from high temperature and high luminosity to low temperature and low luminosity, there seemed to be a second area of the diagram (low temperature and high luminosity) where a significant population of stars existed. Both men concluded that these two groups were related and represented different stages in a star's evolution. Both men also concluded that since the stars with low temperature and high luminosity were rare, this stage of evolution was a more rapid phase. And this is how the H-R Diagram was born.

Scientists recognized that both Hertzsprung and Russell had come upon an aspect of the physical nature of stars that would not be revealed by studying the spectra of stars alone. The work of the women "computers" at Harvard, in conjunction with the works of Annie Cannon, Megnad Saha, and Cecelia Payne-Gaposchkin, were used to expand the H-R Diagram to learn more about the physical nature of stars and to deduce stellar evolution. The details of how these three individuals contributed to the modern understanding of stellar evolution were detailed in chapter 1.

Temperature, Spectral Class, and Color

As discussed earlier in chapter 1, the color a star appears to be is related to its surface temperature. Using a star's color, one can place a star somewhere on the x-axis of the H-R diagram. As discussed in chapter 1 the spectrum of a star is also related to its surface temperature. The work that Annie Cannon did enabled astronomers to determine the temperatures of hundreds of thousands of stars. Further, thanks to the work of Hertzsprung, Russell, and Antonia Maury, stellar spectra can give astronomers information about the size of the star, which tells astronomers where it lies on the H-R Diagram, even without the knowledge of the star's absolute magnitude, luminosity, or distance.

On the H-R Diagram, the horizontal axis can be temperature, spectral type, or color. If the temperature is on the horizontal axis, it goes from left to right, from high to low temperature, or "backwards." The spectral types go in order from left to right in the order Cannon derived: O B A F G K M. (The spectral types and the order of the letters is discussed in detail in chapter 1.) The "colors" of the stars go from left to right, blue to red, in the order the colors appear in the visible part of the **electromagnetic spectrum.** If one examines a rainbow or the spectrum produced by a prism or crystal, the colors go

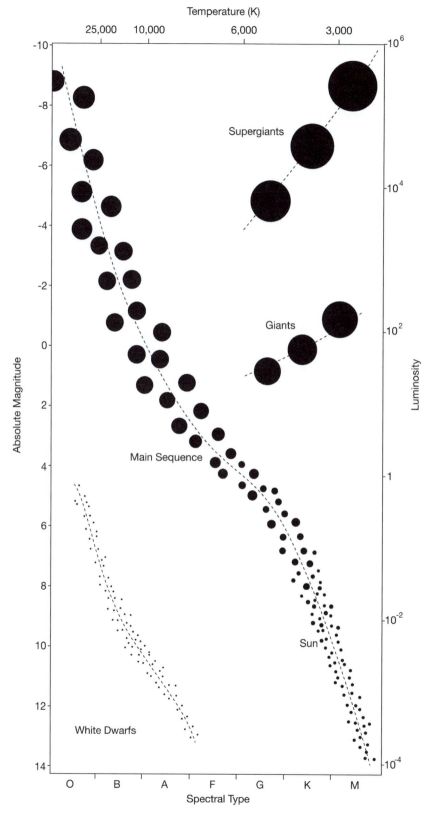

Figure 2.2 The above H-R Diagram shows the four main groupings discovered by plotting the nearest and brightest stars. These groups are (1) the Main Sequence, (2) the Giants, (3) the Supergiants, and (4) the White Dwarfs. Each group is a stage in the evolution of a star. [Jeff Dixon]

(from the blue end to the red end) as follows: violet, indigo, blue, green, yellow, orange, red. The human eye does not distinguish well between violet and indigo, so most humans usually describe the end of the spectrum as "purple" or "blue."

A star's temperature (on its surface) is related to its peak wavelength. This is the wavelength at which most of the star's energy is being radiated. Stars that have their peak wavelengths in the blue end of the electromagnetic spectrum (the peak can be anywhere from the blue end of the visible part of the spectrum to the ultraviolet part of the electromagnetic spectrum) appear bluish. Stars that have their peak wavelengths in the middle of the visible part of the electromagnetic spectrum usually appear white because these stars emit most of their energy all throughout the visible part of the electromagnetic spectrum. When all colors of light are mixed together, the light appears white. Stars with peak wavelengths in the middle-red end of the visible part of the electromagnetic spectrum appear yellow or golden. Finally, stars with peak wavelengths in the red end of the electromagnetic spectrum (the peak can be anywhere from the red end of the visible part of the spectrum to the infrared part of the electromagnetic spectrum) appear red-orange.

On the H-R Diagram, along the x-axis, the temperature goes from hottest to coolest, with temperature increasing to the left. Knowing this and the information given earlier, we can begin to piece together the physical characteristics of a star by its temperature, color, or spectral type. An O star, for example, has a temperature ranging from about 20,000 to 50,000 degrees Kelvin, and will appear bluish-white. On the other hand, a red star must have a spectral type K or M and a surface temperature of 2,000–3,000 K. Our Sun has a surface temperature of about 5,500 K, it appears yellow-white and is a spectral class G star. Knowing one of the three possible quantities that make up the x-axis of the H-R Diagram tells one at least the range of the possible values for the other two quantities. But wait, there's more!

Luminosity and Absolute Magnitude

As discussed above, luminosity and absolute magnitude are related in the same way brightness and apparent magnitude are related. The mathematical relationship between luminosity and absolute magnitude incorporates the logarithmic nature of human vision. The equation is shown below, but a deeper understanding of this equation is beyond the scope of this series.

$$L_*/L_\odot = 2.512^{(4.83-M_*)}$$

In this equation L_* is the luminosity of the star in question, $L_\odot$ is the luminosity of the Sun, and M_* is the absolute magnitude of the star in question. The above equation is valid for measurements using the V (visual) astronomical filter.

Filtering a Star's Light

Astronomers use filters to examine parts of the electromagnetic spectrum using photometers (photon counter) or imagers (film or digital camera), rather than spectrometers. Since a spectrometer spreads out the star's light into its different colors, it takes longer to collect enough light to take a spectrum image than it does to collect enough light to take an image of a star without spreading out its light. The spectrum of a star, however, provides a vast amount of information that is useful to astronomers studying the stars and trying to understand their physical nature. As a compromise, astronomers found that by filtering a star's light into different **bands** they could retain most of the information found by studying a star's spectrum. So, filters were developed to sample parts of the electromagnetic spectrum in the ultraviolet (U), blue (B), visible (V), and red (R), parts of the spectrum. Using these four filters, astronomers can measure the brightness of a star in each band using an imager or a photometer and compare it to the brightness in another band to determine an approximate spectral type and surface temperature.

The absolute magnitude and the luminosity of the star comprise the y-axis of the H-R Diagram. Knowledge of either the absolute magnitude or the luminosity of a star requires some knowledge of the star's distance from Earth. Apparent magnitude can be measured from a simple brightness measurement, but distance is required to determine luminosity or absolute magnitude.

The fact that two pieces of information are needed to put a star on the y-axis of the H-R Diagram means that three pieces of information are needed to place a star on the diagram: temperature (or spectral class or color), apparent magnitude (or brightness), and distance (to get absolute magnitude or luminosity). From this information, it might seem that it would take a long time to be able to put enough stars on the H-R Diagram to begin to see any trends; however, because of the work so many astronomers were doing in the early 20th century, there were already plenty of stars with sufficient data to populate the diagram and learn much about the physical nature of stars.

MAIN SEQUENCE STARS

Upon populating the H-R Diagram with stars, astronomers noticed some very interesting features. The first, and perhaps most important feature astronomers noticed, was that about 90 percent of the stars plotted formed a line that goes from the upper left corner to the lower right corner of the diagram. Since most stars fell on this line, it was called the "main sequence." The Sun, it turns out, is a main sequence star. Main sequence stars span the full possible range in brightness, and temperature; however, using the Stefan-Boltzmann relation, we can see that main sequence stars span a small range of sizes from 10 times smaller than our Sun to 10 times larger than our Sun.

Because so many of the stars plotted fall on this line, astronomers surmised that the main sequence line must be a very important piece of information about how stars exist. However, it wasn't until astronomers started to look at the H-R Diagrams of clusters of stars that they were able to piece together the meaning of the main sequence. This discovery will be discussed in greater detail in chapter 3. The main sequence of stars follows a straight-forward mathematical relationship that describes the existences of the stars upon it. Main sequence stars with high luminosities have high temperatures and are larger and more massive than main sequence stars with low luminosities (which have low temperatures are smaller and less massive).

What astronomers now know is that the main sequence has temporal (time) significance. It describes not just the majority of stars, but the majority of a star's existence. All stars begin their existence as stars on the main sequence. Astronomers also now know that 90 percent of the existence of a star will be on the main sequence. That is to say once a star leaves the main sequence (once it changes temperature and size), 90 percent of that star's existence has been completed and the star will soon cease to exist.

Luckily for us, our Sun is still on the main sequence, indicating that it still has a lot of time left before it changes temperature and size and begins its journey to its end. Astronomers learned all this from the H-R Diagram when they started to look at how clusters of stars appear on it. How astronomers figured all this out and more detail about what we now know about stars will be discussed in chapter 3.

THE GIANTS

In addition to the main sequence, three other main groupings of stars on the H-R Diagram were noticed. One such grouping was named the giant stars. Because these stars are located in the upper middle part of the H-R Diagram, we know they are fairly luminous, and because they are on the right-hand side of the H-R Diagram, we know they must have low surface temperatures. So, their location on the H-R Diagram tells us these stars are luminous and have a very low surface temperature.

Using the Stefan-Boltzmann equation, we can conclude that these stars must be very large to give off so much energy with such low surface temperatures. In fact, the giant stars range from 10 to 100 times larger than their main sequence counterparts with the same luminosities. Since these stars are on the right-hand side of the H-R Diagram, and therefore are red in color, these stars are commonly known as red giants.

Again, as with the main sequence stars, this grouping of stars seemed important (especially since there were no stars on the H-R Diagram outside of these four main groups). The number of stars in the giant group was far fewer than the number of stars in the main sequence, but stars appear in this group, so it must be a significant phase in a star's existence.

What astronomers know now is that the red giant phase is a phase that many stars go through when they leave the main sequence. Since all stars that have evolved past the main sequence have been through this phase, this is the second most important group on the H-R Diagram. This phase does not last long, however, so the stars in this group are not nearly as many as the stars in the main sequence. Again, all this is information we know now because astronomers spent so much time studying how clusters of stars appear on the H-R Diagram.

THE SUPERGIANTS

A much smaller group of stars is located in the upper part of the H-R Diagram spanning from the upper main sequence through the right side of the diagram. Because these stars are in the upper part of the H-R Diagram, we know that they are very luminous stars. Because they span from the upper main sequence to the right side of the H-R Diagram, we know they have lower surface temperatures than their main sequence counterparts. Again, using the Stefan-Boltzmann Law, we can conclude that these stars must be larger in size to produce such high luminosities with lower surface temperatures. These stars are more luminous and can be much larger than the giants, so they were named supergiants.

Supergiant stars are extremely rare, although there are a few in the neighborhood of the Sun. These stars range from 10 to 1,000 times larger than their main sequence counterparts with the same luminosities. These stars distinguish themselves from the giant stars by their luminosities. The supergiant stars are the most luminous stars, about 10 to 100 times more luminous than the giant stars. Another significant difference between the giant and supergiant groups is that there are supergiant stars with surface temperatures much higher than the giant stars. This means there are both red and blue supergiant stars, whereas there are only red giant stars.

Astronomers now know that supergiant stars were once the most luminous and most massive main sequence stars. This phase of a massive star's existence is very short, so very few stars exist in the supergiant group. However, because these stars are so luminous, astronomers have probably observed all that exist in the neighborhood of the Sun.

Astronomers cannot say this about populations of stars with lower luminosities. What makes a star easy to observe is its brightness. So, the most complete samples of stars will include nearby stars with medium luminosities and all stars with high luminosities. That is to say, our understanding of stars is limited by our ability to capture their light for study. So, the brightest stars are the ones about which we have the most complete information.

THE WHITE DWARFS

The last of the main groupings of stars found on the H-R Diagram is a group found in the lower left portion of the H-R Diagram. This group looks different from the previous two groups because it forms a narrow band parallel to the main sequence. Because this group is in the lower part of the H-R Diagram, we know these stars are not very luminous. Because this group is on the left part of the H-R Diagram, we know these stars have very high surface temperatures.

If we use the Stefan-Boltzmann Law, we can easily see that these stars must be very small to give off so little light at such high surface temperatures. This group of stars was called the white dwarf stars. Because of their high surface temperatures, these stars appear white. White dwarf stars are about 10 to 100 times smaller than their main sequence counterparts of the same luminosity. Astronomers now know that white dwarfs are the final stage of existence for stars that go through the giant phase.

RECOMMENDED READINGS

Asimov, Isaac. *Asimov on Astronomy.* New York: Bonanza Books, 1988.

Bennett, Jeffrey D., Megan Donahue, Nicholas Schneider, and Mark Voit. *The Cosmic Perspective.* 5th ed. San Francisco: Benjamin Cummings, 2007.

DeGrasse Tyson, Neil, Charles Tsun-Chu Liu, and Robert Irion. *One Universe: At Home in the Cosmos.* Washington, DC: Joseph Henry Press, 1999.

Freedman, Roger, and William J. Kaufmann III. *Universe.* 8th ed. New York: W.H. Freeman Company, 2008.

Hawking, Stephen. *On the Shoulders of Giants.* Philadelphia: Running Press, 2002.

Jastrow, Robert. *Red Giants and White Dwarfs.* New York: W.W. Norton and Company, 1990.

Seeds, Michael A. *Astronomy: The Solar System and Beyond.* 5th ed. Pacific Grove, CA: Brooks Cole, 2006.

3

Putting Together the Puzzle of Stellar Evolution

In the last chapter, intrinsic properties of stars and the H-R Diagram were introduced. The H-R Diagram, in its inception, revealed the first clues to understanding stellar evolution. Just from plotting the nearest stars and the brightest stars, astronomers learned that stars fall into one of four general categories. Astronomers identified the most common category as the **main sequence.** And, early on, astronomers surmised that the second most common group (the giants) must be a stage in the evolution of stars, albeit a shorter stage than the main sequence stage.

Currently we know much more about stellar evolution and even have created models of what is happening in their interiors and how the light we see is formed and transported from the core to the atmosphere of the star. How did astronomers figure all this out? How can they be so sure their models are correct? These are the questions this chapter will attempt to answer.

STAR CLUSTERS

The key to understanding stellar evolution came with the study of star clusters. In the early 20th century, astronomers began to realize that some of the **nebulae** they could observe were not just clouds, but stellar nurseries. When astronomers realized that stars could form in groups, or clusters, they realized the value in studying star clusters. Such clusters of stars would be approximately the same age and would be located in the same part of space (meaning that they would all be the same distance away from the Sun).

Even before the realization that some nebulae were stellar nurseries, clusters of stars were well-known and catalogued. (What was yet to be known about the nebulae was that stars formed this way—already in clusters, and from the same materials.) Astronomers had identified two basic types of star clusters: **globular clusters** and **open clusters.** Globular clusters are dense clusters of stars in a nearly spherical formation. Open clusters are loose, shapeless associations of stars.

Globular clusters contain about 10,000 to 1,000,000 stars. Usually little or no gas or dust is found in a globular cluster. Globular clusters are gravitationally bound systems of stars. These objects are relatively small, compared to a galaxy. Where a galaxy, like the Milky Way, is about 30,000 pc across, a globular cluster can be anywhere from about 3–60 pc across. (Compared to our solar system, this is very large. Our solar system is only about 0.0004 pc across!) So many stars in so little space means the stars are very close to one another. Globular clusters are also mostly found above and below the plane of the Milky Way. That is, if the Milky Way is shaped something like a pancake, the globular clusters are like the syrup or the butter. They are very near to the galaxy, but mostly they are not in the same part of the galaxy that our Sun is in (the pancake part).

Open clusters (sometimes called galactic clusters because they are mostly located in the disk of the Milky Way, and so appear in the galaxy) are loose associations of stars. It is difficult to accurately define limits to the number of stars or sizes of open clusters because even the most well-studied open clusters may have members that have not yet been identified. Because of their

Globular Clusters and the Shape of the Milky Way

Harlow Shapley used globular clusters to deduce the size of the galaxy and our position within it. Before Shapley, astronomers had concluded that the solar system was located at the center of the galaxy. This determination was based on star counts. Astronomers counted how many stars they could see in different directions. The numbers indicated that the solar system was located in the center of a disk-shaped galaxy.

Shapley built on the work of Caroline and William Herschel, who discovered the presence of dark dust clouds. Shapley concluded that the dark clouds might affect the star counts that earlier astronomers used. He thought it would be better to use globular clusters to learn about the shape of the Milky Way, since they were not in the same place the dark dust clouds were located. Because the globular clusters are not in the plane of the galaxy, but are still distributed throughout the galaxy fairly evenly, they are a good way to learn about the size and scope of the galaxy and to determine our position within the galaxy.

This idea is similar to the idea of using a lighthouse to identify landfall on a foggy bay. The fog is low and close to the water, whereas the lighthouse is tall and high above the layer of fog. The boat on the water can see the lighthouse because the light source is above the fog and the person in the boat is not looking through as much of the fog to see the light. Shapley thought that the globular clusters were like the lighthouses and the disk of the galaxy is like the fog.

location within the plane of the Milky Way, open clusters are more difficult to distinguish. Sometimes a few prominent members are obvious (like in the case of the Pleiades), but the other members of the group may be more difficult to identify. The plane of the galaxy is full of stars, and identifying a cluster within the plane is difficult to do, unless you know each star's distance from Earth.

Unfortunately, for some stars, measuring distance is just not possible. Some stars are too far away to measure their distances using parallax. While there are other ways to measure distance, they require that the star be a special type of variable star (discussed in much greater detail in chapter 4).

....

The Period-Luminosity Law

Some special kinds of variable stars (Cepheids and RR Lyrae, for example) follow a period-luminosity law. This means that the period of variation that the star has (the time it takes for the star to go from bright to faint to bright again) is mathematically related to its peak luminosity or absolute magnitude. So, an astronomer can measure the period and the maximum brightness (apparent magnitude) of one of these types of variable stars, and from those data, she could calculate the distance to the star.

....

What is special about clusters is that all the stars within a cluster were formed from the same cloud at approximately the same time, so that they are all the *same age*. Also, because they are all in the same part of space, so they are all the *same distance* from Earth. These small details were the key to how astronomers were able to figure out stellar evolution.

Color-Magnitude Diagrams

Since the distance to most stars in clusters are not known, it is not possible to put them on an H-R diagram; however, since the axes of the H-R diagram are temperature and luminosity, astronomers can use other information from stars in clusters to plot them on a similar diagram. Recall that color is similar to temperature, so astronomers could, conceivably, use color, rather than temperature or spectral type. In addition, recall that the absolute magnitude of a star is its apparent magnitude, if it were at a distance of 10 pc. Since all stars in a cluster can be considered to be at the same distance from Earth, all the apparent magnitudes of the stars will be offset from the absolute magnitudes by the same amount (proportional to their actual distance from Earth, divided by 10 pc). So, astronomers plot a star's color on the x-axis (since color corresponds to temperature) and a star's apparent magnitude on the y-axis (since apparent magnitudes of stars that are all at the same distance from Earth correspond to a star's absolute magnitude). These diagrams are called, therefore, color-magnitude diagrams.

Measuring apparent magnitude is, perhaps, easy to understand. It is a simple measure of a star's brightness. This is a process called photometry. A star's image is taken with either a photometer or an electronic imager, similar to a digital camera. The star's brightness can then be measured by either a calculation based on the parameters of the instrument (photometer or electronic imager) or by comparison to other stars with known brightnesses on the electronic image.

Measuring a star's color is a bit more complicated. First, the star's light must be sampled in at least two different places in the electromagnetic spectrum. Astronomers realized that obtaining spectra of every star in the sky with enough clarity to assign it a spectral class would take far too long, even if one could take hundreds or thousands of spectra at one time. So, for ease and for simplicity, astronomers developed a way to determine a star's color (and therefore spectral class) by dividing up the visible part of the electromagnetic spectrum into four parts and sampling these parts to determine its spectral class. Each part of the spectrum is sampled using a filter that allows a small fraction of the visible light through and blocks all other wavelengths. These filters are called U, B, V, and R for the ultraviolet, blue, visual and red parts of the visible electromagnetic spectrum. Since all stars give off different amounts of light across the visible part of the electromagnetic spectrum, some information about spectral type can be determined simply by determining how much light a star is giving off in one part of the spectrum relative to some other part of the spectrum.

A star's color is a comparison of the brightness of the star in any two of these four filters. Usually, astronomers use the B and V filters since most stars are brightest in these parts of the electromagnetic spectrum. Color is a number that is the difference between the apparent magnitude of a star in two different filters. A blue star would have a B-V color that is smaller than that of a red star. This is because a blue star will be brighter in B than in V, and a red star will be brighter in V than in B. And, since the magnitude system is backwards, bigger numbers mean fainter stars, a blue star will have a bigger V and a smaller B, so B-V will be a negative number. A red star, on the other hand, will have a smaller V and a bigger B, so its B-V color will be a positive number.

Amazingly, what astronomers found when they studied their color-magnitude diagrams of open and globular clusters was that this method not only gave the clues to stellar evolution, but it allowed them to determine the distances to the clusters. The color-magnitude diagrams of clusters look like H-R Diagrams. There is an obvious main sequence and in some, there are giant and supergiant stars. Additionally, in color-magnitude diagrams these groups (main sequence, giant, and supergiant) are not discrete—they are connected to one another, indicating that stars change in a gradual way from the main sequence to each of these other groups.

When astronomers saw that color-magnitude diagrams of the stars in clusters mimicked the H-R diagrams of stars in the solar neighborhood, they realized that the stars in clusters could reveal the story of stellar evolution.

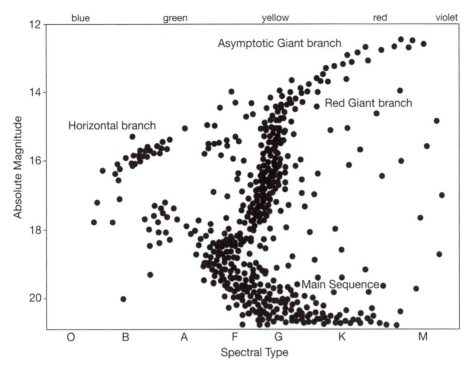

Figure 3.1 Above is a color-magnitude diagram of a cluster of stars. The y-axis is the apparent magnitude and the x-axis is color (the difference between the apparent magnitude of the star in two different filters). The similarity of this diagram to the H-R Diagram is what prompted astronomers to piece together stellar evolution. It was apparent that clusters of stars were formed simultaneously and contained stars at various stages in their evolution. The sequence of stellar evolution could be pieced together by studying different clusters.. [Jeff Dixon]

In the color-magnitude diagrams of clusters of stars, the main sequence is not always complete. In fact, it turns out, the older the cluster, the shorter the remaining main sequence. Also, the main sequence disappears always from the upper left hand side (the hottest, most luminous, and most massive stars), indicating that these stars exist for the shortest length of time.

Looking at these color-magnitude diagrams, it is clear that stars leave the main sequence by increasing in luminosity and decreasing in surface temperature. That is, the stars change position on the color-magnitude diagram as they age. Stars move from the main sequence up and to the right to the giant and/or supergiant groups. They move through a region now known as the **red giant branch,** and some will move along the horizontal branch, while some move through the asymptotic branch. This level of detail about how stars evolve is only possible to know by exploiting the laboratories that are star clusters.

In clusters of stars, all born at essentially one instant in time and all in one location in space, stars are evolving. It turns out that how a star evolves depends mostly on its mass. Since the stars in a cluster are formed with a range of masses it is possible to observe the evolution of stars with a wide range of masses. And, since they started fusing hydrogen at approximately the same time, they are all at different stages in their evolution. For this reason, astronomers see many stars at points in their evolution that are very short-lived. Because these stages are short-lived (lasting years or thousands of years), it is not likely that many stars in a random sample will be observed in this stage of evolution. In a cluster of stars, however, at any given moment, there should be stars in almost every stage of evolution.

EVOLUTIONARY TRACKS

Using the information gained by studying many globular clusters and open clusters, astronomers were able to deduce the evolutionary sequence for stars. Furthermore, from the data they gathered, they could trace evolutionary tracks on the H-R Diagram showing how a star changes in physical appearance over the length of its existence.

The first clue to stellar evolution is that the main sequence disappears from the upper left. This means that the most massive, most luminous, hottest stars have the shortest existence on the main sequence. In fact, the data show that the older a cluster is, the more diminished the main sequence is. Astronomers used this information to create models that explained these data. Knowing that main sequence stars produce energy through hydrogen fusion, making some assumptions about dominant energy transfer processes and the density of material at different points within the star, astronomers can determine the length of time a star of a given mass will remain on the main sequence. With this knowledge, the age of a cluster can be determined by observing the location on the main sequence of the highest mass star (the **turn-off point**) and then calculating the mass of that star and the length of time it can fuse hydrogen. This length of time is the same as the age of the cluster,

For example, take the well-known open cluster the Pleiades. This open cluster is also known as the "Seven Sisters" and is found in the constellation Taurus. While most people can see six or seven stars in this cluster, there are actually more than 600 stars considered to be members of this cluster.

To determine both the distance to the object and the age of the cluster, a color-magnitude diagram of the Pleiades is superimposed on an H-R Diagram. The main sequence is then lined up on both diagrams so that the main sequence of Pleiades aligns to the main sequence on the H-R Diagram.

Then, to determine the distance to the cluster, the vertical axis on the color-magnitude diagram is matched to the vertical axis on the H-R Diagram. Recall that the vertical axis on the H-R Diagram is absolute magnitude. The vertical axis on the color-magnitude diagram is apparent magnitude. So, when the two main sequences are matched up, the difference between the absolute and apparent magnitude of any star on the diagram can be determined. Hence, the distance can be calculated using the following equation:

$$m - M = 5 \log d - 5$$

where M is the absolute magnitude, m is the apparent magnitude, and d is the distance in parsecs.

Now, to determine the age of the cluster is a little bit more complicated. The turnoff point can be noted on the color-magnitude diagram before aligning it to the H-R Diagram, but it is still necessary to know the absolute magnitude of the turnoff point. The turnoff point is the point at which the main sequence appears to turn off. For the Pleiades, the turnoff point on a color-magnitude diagram is at about 6.5 magnitudes. The absolute magnitude of this turnoff point is about 1.

Now comes the tricky part. We have to interpret, from this absolute magnitude, the age of the cluster. The way this is done is by understanding how stars produce energy while on the main sequence and what process occurs to cause them to move from the main sequence to the red giant portion of the H-R Diagram. We know that this phenomenon is due to the star's internal energy source switching from hydrogen fusion to helium fusion. To get a good idea of how long a star at a particular point on the main sequence will stay on the main sequence, we can use what we know about the mass of the star and calculate how long it will be until it must switch from hydrogen fusion to helium fusion in its core. For a star on the main sequence with an absolute magnitude of 1, this length of time is about 100 to 150 million years, depending on some assumptions that must be made. The details of how to make this calculation are discussed below.

In addition to being able to determine the age of a cluster, studying the color-magnitude diagrams of clusters also allowed astronomers to fill in gaps in their picture of stellar evolution that existed due to lack of evidence from field stars (stars not in clusters). Astronomers learned *how* stars change as they go from fusing hydrogen on the main sequence to fusing other elements as they move along the red giant branch, become red giants, and so on. Astronomers discovered the red giant branch, the **asymptotic branch,** the **horizontal branch,** and the **instability strip** among other things.

The red giant branch is the first stage of a star's existence after the main sequence. The most massive stars in a cluster start on this stage of evolution before the less massive stars. In the red giant branch stage of a star's life, the star is still fusing hydrogen, but not in the star's core. Hydrogen fusion is occurring in a shell surrounding the core. As the fusion occurs further from the center of the star, the outer parts of the star are expanding and the surface temperature of the star is dropping. So the star's luminosity is increasing, while its temperature is decreasing; this set of circumstances indicate that the star is moving, as seen on the H-R Diagram, up and to the right of the main sequence.

At the tip of the red giant branch is where the transition from hydrogen shell fusion to core helium fusion happens for stars massive enough to achieve this. The core of the star collapses a little, increasing its density, and heats up to a much higher temperature and helium fusion begins. This new energy source is in the core and not in a shell surrounding the core, so the star shrinks back down to a smaller size (still larger than it was as a main sequence star).

Stars that are stable and fusing helium in their cores enter the horizontal branch increasing in temperature and decreasing in luminosity (moving down and to the left on the H-R Diagram). As this occurs, some stars enter the instability strip and become pulsating variable stars (discussed in great detail in chapter 4).

If a star is massive enough to be able to fuse carbon after it runs out of helium fuel in its core and shell, it will first follow the asymptotic giant branch (which runs parallel to the red giant branch). The asymptotic giant branch will take the star back to the giant group again, decreasing the surface temperature and increasing in luminosity as it expands (moving up and to the left on the H-R Diagram). Following this trip back to the giant group, shell hydrogen and helium fusion begin, the star's core collapses still more, increasing its density and heating up to ignite fusion of carbon. Again, the star stabilizes, igniting carbon fusion in the core, this time; the star shrinks down (still remaining larger than the last time it shrunk) and heats up the outer layer (moving downward and to the left on the H-R Diagram).

Each new fuel the star consumes allows the star to remain stable for a shorter and shorter period of time. This is presumed by the small number of stars found in this part of the H-R Diagrams of clusters. A small number of stars in a region implies that the region is not occupied by stars for a long time. (This is similar to the conclusion that because 90 percent of all stars are on the main sequence, this is the state in which most stars exist most of the time.)

From what is understood about energy production of fusion in terms of atoms and atomic structure, astronomers know that stars can fuse elements up to iron; fusing iron, however, requires energy to be added. Therefore, if stars fuse iron, it will not help to keep the star in balance. It will cause the star to collapse because without producing energy, there is no outward pressure to balance gravitational collapse and the star will simply collapse.

From studying and trying to reproduce the fusion of elements in laboratories on Earth, astronomers know that the fusion of heavy elements (elements heavier than helium) requires extremely high densities of material and extremely high temperatures. These conditions could, theoretically, occur in the cores of stars, but do not likely occur in the outer parts of stars. So, astronomers can calculate the amount of material that is needed so that the star would be dense enough and hot enough to fuse heavy elements. The amount of material available for fusion gets smaller and smaller with each switch to a new fuel, which explains why each phase of the star's existence is shorter and shorter. It is estimated that stars that fuse silicon (which produces nickel, which rapidly decays into iron) do so for only about a day (this time varies with the mass of the star, but not by much). Given that stars can exist for millions and billions of years, fusing an element to sustain itself for only one day is really a star's final gasp.

PRE-MAIN SEQUENCE STARS

So, stars evolve off the main sequence, and looking at clusters tells astronomers much about how a star changes as it evolves. Studying very young star clusters also reveals how stars *become* main sequence stars. In some color-magnitude diagrams of very young clusters of stars, astronomers saw strange groupings of objects that appeared very red. As astronomers began to observe clusters using infrared detectors (which became more and more common in the 1980s and 1990s), they noticed that in the youngest clusters, there were a large number of these objects that could be plotted on color-magnitude diagrams.

These strange objects turned out to be **pre-main sequence stars** (often referred to as PMS stars). These objects are actually stars that are still in the process of clearing away the debris from the clouds from which they were formed. Some of these objects are also known as **T Tauri** stars.

••

T Tauri Stars

T Tauri stars are named after the first such object observed, an object in the constellation Taurus. By the letter T designation, it is apparent that this object is one of the fainter stars in the constellation Taurus. This star was discovered in 1852 by an astronomer named John Hind. At the time T Tauri was noteworthy because it was an irregular variable star. A variable star is one that varies in brightness. Astronomers were just beginning to realize that most stars that vary in brightness do so in a regular and predictable fashion. This star, however, is one that is not regular. It turns out that the reason its variation in brightness is not regular is because its brightness is varying for reasons that are very different from those stars which vary in brightness in a regular and predictable way. Regularly varying stars vary their brightness, for the most part, by expanding and contracting. T Tauri stars are thought to vary their brightness due to instabilities in their surrounding **accretion disks.**

T Tauri stars have spectral types G to M, so they appear to be early stages in evolution of low-mass main sequence stars, like our Sun. These stars are thought to go through a phase in their evolution where a disk of debris (from the nebula from which they formed) is surrounding the star. Some material from the disk is accreted onto the star and may be spewed back into the outer parts of the disk. Hotspots may occur in the disks of these stars and planets may form in the disks of these stars. Much of the current research in stellar evolution and planet formation involves studying these kinds of stars to better understand how they evolve and how our Sun and solar system may have come to be. Still, much is not known, but models and observations are narrowing the range of possibilities for how stars like our Sun may evolve.

••

There are other populations of stars that are known to be PMS stars, but the most interesting ones to astronomers right now, are the T Tauri stars. This is probably because they offer so much insight into how our solar system formed and how other solar systems may form. In astronomy, what is interesting to astronomers is where the research will be. Therefore, very little is known about higher-mass PMS stars. Recently, x-ray observations may be revealing information about intermediate- to high-mass PMS stars. It is still uncertain, however, as to whether these objects are high-mass PMS stars.

Pre-Main Sequence Evolution

From studying the youngest clusters of stars, astronomers have been able to deduce pre-main sequence evolutionary tracks. These tracks are not entirely supported by observations as the post-main sequence evolutionary tracks are. Pre-main sequence evolutionary tracks are more highly dependent on a small number of objects and extrapolation based on physics and what little is known about the interior structure of stars.

Astronomers began to try to piece together the information they had into a coherent picture of how a star comes to be. For stars less massive than our Sun, the picture was fairly complete by observation alone, but for stars like our Sun or more massive than our Sun, what happens before the objects become main sequence stars is more mysterious. One could fairly easily use mathematical models based on the low-mass stars and extrapolate to higher-mass stars, however. And this is what has been done.

Hayashi pre-main sequence evolutionary tracks use both mathematical and observational data to describe what happens to objects that are less than one half the mass of our Sun before they become main sequence stars. Heyney pre-main sequence evolutionary tracks use mostly mathematical and some observational data to describe what happens to objects that are more than one half the mass of our Sun before they become main sequence stars. Both models imply a predictable sequence of events and a calculable period of time for objects to become main sequence stars.

STAR LIFETIMES

As we learned above, stars exist for varying amounts of time. The amount of time that a star exists depends on its mass. The amount of time that a star can exist has everything to do with how and why stars exist in the first place. The mechanism for producing energy inside a star is behind all this. If the star produces energy, it exists. Otherwise, something will be out of balance and the star will cease to exist.

In this section we will learn, in detail, how a star produces energy, how this governs a star's existence on the main sequence, and, for low-mass stars, how this governs the rest of a star's existence. The entire sequence of stages of a star's existence (for low-mass stars) will be outlined in detail here.

As you already know, this evolutionary sequence was determined by studying clusters of stars of varying ages and deducing the evolutionary sequence for stars of all masses. How this sequence is explained by the physics of the interior structure and mechanisms for energy production in stars will be revealed in this section.

THERMONUCLEAR FUSION OF HYDROGEN

The first and most important process to understand is thermonuclear fusion of hydrogen. Often, astronomers refer to this process as "hydrogen burning," but this is not truly "burning." Burning is a chemical process that involves only electrons. Thermonuclear fusion of hydrogen involves no electrons, just protons and neutrons. Inside the cores of stars, hydrogen atoms are protons. The electrons are stripped from the atoms in the dense cores of stars.

Thermonuclear fusion of hydrogen occurs only under the conditions of extremely high density and extremely high temperature. The density of the core of our Sun (which is currently fusing hydrogen as its energy source) is about 150 grams per cubic centimeter, which is 150 times the density of water and a little more than 10 times the density of lead. Compared to the most dense material on Earth (osmium or iridium), the core of the Sun is about seven times more dense. The temperature of the core of our Sun is about 15 million degrees Kelvin (or about 27 million degrees Fahrenheit).

These conditions are necessary for hydrogen fusion to occur. Because hydrogen atoms are just protons, they resist being pushed together (since they are both positively charged). So, they must be moving very fast to collide and "stick" together (hence the high temperature), and they must be very close together to have a high enough probability of colliding with one another for these collisions to occur.

The process of fusing hydrogen is not as simple as just taking two hydrogen atoms and pushing them together to make a helium atom. There are some unexpected transformations that have to occur. Thermonuclear fusion of hydrogen happens in three steps. Since two of the steps are interdependent and must occur twice for the last step to happen, the number of steps is actually five.

In step one, two hydrogen nuclei (two protons) collide. As a result something unexpected happens. One proton turns itself into a neutron. This process is called proton decay or **beta decay.** This process was named and discovered in the study of radioactivity. Some radioactive substances undergo beta decay where a neutron decays into a proton emitting also an electron and an anti-neutrino. So, the first step of thermonuclear hydrogen fusion that occurs in the cores of main sequence stars involves beta decay. The nuclear reaction can be written as follows:

$$p + p \rightarrow D + e^+ + \nu_e$$

In this equation, p is for proton; D is for deuterium (also known as "heavy hydrogen"), which consists of one proton and one neutron; e^+ is the symbol for a positron (anti-electron); and ν_e is the symbol for an electron **neutrino.** Neutrinos are weakly interacting particles that have almost no mass. They are associated with leptons (the family of particles of which electrons are a member), so there are three kinds of neutrinos: electron, muon, and tauon neutrinos.

In step two, the deuterium nucleus (a proton-neutron pair) interacts with another proton and forms the unstable form of helium known as helium-3. This is a nucleus containing two protons and one neutron. In this part of the reaction, a gamma photon is released because the energy of the nuclear bond that holds together helium-3 is less than the energy of the nuclear bond that holds together deuterium. The nuclear reaction described in step two can be written as follows:

$$D + p \rightarrow He^3 + \gamma$$

where D is for deuterium, p is for proton, He^3 is for helium-3, and γ is for the gamma ray radiation.

To accomplish step three, steps one and two must have been completed again so that there will be another helium-3 nucleus to interact with. In step three, two helium-3 nuclei interact and form one stable helium-4 nucleus and two protons. The nuclear reaction for this step can be written as follows:

$$He^3 + He^3 \rightarrow He^4 + p + p$$

where He^3 is for helium-3, He^4 is for helium-4, and p is for proton.

So, if we put together all three steps (including steps one and two, two times each) the full reaction could be written as follows:

$$6p \rightarrow He^4 + 2p + 2\gamma$$

The equation could be further reduced to:

$$4p \rightarrow He^4 + 2\gamma$$

It is from this form of the equation that hydrogen fusion is usually taught. Four protons come together and make a helium atom. Some energy is given off because the mass of four hydrogen atoms is greater than the mass of one helium atom. This is how astronomers can calculate the amount of energy given off by this reaction. A hydrogen atom has an atomic mass of about 1.008 amu (atomic mass units = 1.66×10^{-27} kg), four times this would be 4.032 amu. A helium atom has a mass of 4.0026 amu. Four hydrogen atoms have about 0.03 amu more than one helium atom. This very small amount of "mass" is converted into energy using the famous equation, $E = mc^2$. If the proper conversions are done, the missing mass converted to energy is equal to the energy of two gamma ray photons. So, even though the above equation leaves out much of what happens during the process, it is sufficient to explain exactly what happens in the process of thermonuclear fusion of hydrogen into helium.

MAIN SEQUENCE LIFETIMES

While a star is on the main sequence, it is stable. This stability is due to a balance between the outward pressure due the material in the core and the gravitational pull of the mass of the star that causes the star to collapse. There is both radiation pressure and gas pressure at work in the case of a main sequence star.

As a star collapses within the cloud from which it forms, there is some resistance to gravitational collapse due to the fact that it takes some energy loss to the system to compress gas. As the density of the gas increases, so does the resistance to gravitational collapse. Additionally, once the temperature and density of the gas are sufficient to allow for thermonuclear fusion to occur, a radiation pressure will appear within the core and eventually throughout the star. The gamma ray radiation released by the reactions in the core will propagate toward the less dense material and eventually out of the star. The migration of this radiation outward constitutes another outward "force" that acts to resist gravitational collapse. This force is far more important to the processes and physical conditions within the stellar core than in the rest of the star.

To understand how to calculate a main sequence lifetime for a star, we must first realize that not all of the star is made up of hydrogen to begin with. Only about 90 percent of a star (by mass) is hydrogen. The rest is helium with a small amount of mass due to heavier elements. When we think about the star by composition (number of atoms) rather than mass, the picture is a little different. From the point of view of estimating a star's ability to fuse hydrogen for a long time, the percent of the star that is hydrogen by number of atoms is more telling than the percent of the star that is hydrogen by mass. By number of atoms, hydrogen comprises only about 75 percent of a star. The percent by number of atoms of helium is about 24 percent. This means that stars begin with about three hydrogen atoms for every helium atom. So, the core of the star already contains some helium before hydrogen fusion even begins.

The second factor that changes how long a star can fuse hydrogen in its core is what kind of energy transfer processes are going on inside the star.

Every star contains both radiative and convective energy transfer regions. The important difference between the two modes of energy transfer is that **radiative transfer** regions do not move the material of the star, but convective ones do. This means that stars with large **convective transfer** regions in their interiors may be able to fuse hydrogen longer than stars without large convective transfer regions because those with large convective transfer regions can move more "fuel" to the "fire." That is, when the core becomes depleted of hydrogen, a star that has a large convective transfer region near the core can move more hydrogen into the core, thereby allowing the core to fuse hydrogen for a longer period of time.

Interestingly, both extremely low- and extremely high-mass stars contain convective transfer regions near the core. In high-mass stars about halfway from the core to the surface, the energy transfer method goes to pure radiative transfer, so about half the star's material will never be able to reach the core. Intermediate-mass stars, like the Sun, have large radiative transfer regions near the core and transition to convective transfer regions at the outer layers. In these stars, only the material that starts in the core will get fused. In low-mass stars, the radiative transfer zone occurs only at the star's surface. So all the material in these stars will reach the core.

Another factor is the fusion process. In stars like the Sun and in lower-mass stars, the fusion process is that described above. In high-mass stars, however, the cores are generally hotter than in intermediate- and low-mass stars. For this reason, another process *in addition* to the one described above is used. That process is called the "CNO cycle" because it uses a carbon atom as a catalyst to quickly use up hydrogen and make helium. In this cycle, a carbon-12 atom fuses with a proton to make a nitrogen-13 atom, which beta decays to a carbon-13 atom. The carbon-13 atom fuses with a proton to make a nitrogen-14 atom. The nitrogen-14 atom fuses with another proton to make an oxygen-15 atom, which beta decays into a nitrogen-15 atom. Another proton fuses with the nitrogen-15 atom making an oxygen-16 atom, which emits a helium atom to become a carbon-12 atom again. The rates for these reactions are much faster than the rates for the proton-proton chain, and so these stars will use up their hydrogen and create a helium core much more quickly than an intermediate-mass star. It is clear how the convective transfer region occurring so close to the core is helpful in accelerating the fusion reactions since it feeds the core with both hydrogen and carbon.

Taking all this information into consideration and knowing the reaction rates for each reaction that occurs in the cores of main sequence stars, we are able to determine a star's time on the main sequence using the following formula:

$$T = 10^{10} / M^{3.5}$$

where T is the time a star will be a main sequence star in years and M is the star's mass in units of the Sun's mass (or solar masses). ($M^{3.5}$ is the same, mathematically, as saying the square root of *M* raised to the seventh power.) This equation says that the more massive a star is, the shorter its existence on the main sequence will be and vice versa.

LOW-MASS STELLAR EVOLUTION

Already it is clear that what happens to a star must be somewhat tied to the mass of the star. We have already seen that high-mass stars spend less time on the main sequence than lower-mass stars because different processes are going on inside them and because the conditions (temperature, pressure, available atoms) in their cores are different. They have different ways of distributing material to the core and different processes that go on inside the core.

Intermediate- and low-mass stars have more in common, but there are still some differences. In this section of this chapter we will talk about low-mass stellar evolution. This route of evolution will be descriptive of what happens to stars with masses less than eight times the mass of our Sun (but greater than 0.4 times the mass of our Sun). For stars more massive than that, a different route of evolution will be followed. That route will be discussed in the next section of this chapter. For stars less massive than 0.4 solar masses, astronomers don't know what happens. The first of these stars that formed in the early universe have not yet stopped fusing hydrogen, so there is no observational evidence of what they will become when they finish fusing hydrogen.

When Hydrogen Fusion Ends

The event that signals a change for a low-mass main sequence star is the end of hydrogen fusion in the core of the star. This ceases because there is insufficient hydrogen to fuse into helium. Actually, the amount of hydrogen to helium (by number of atoms) gets as low as 1 to 9 (that is 10% hydrogen by number and 90% helium by number).

Because there is no significant amount of fusion occurring (there is, of course, some fusion occurring, but it is not happening at an appreciable rate), the core begins to contract. Gravitational collapse of the core sets in because of the lack of radiation pressure from the diminished number of escaping gamma ray photons. There is still gas pressure to resist the collapse, though, so as long as the mass of the core is sufficiently high, it will not completely collapse.

As the core collapses, the temperature of the core will rise and so will the density of the gas within the core. The outer part of the star is doing something completely different. Fusion does not cease or diminish at all in a thin shell surrounding the core. As the core collapses, it takes some of the material from the outer part of the star with it. That new material from the outer part of the star condenses and gets hot enough to fuse hydrogen. That material from outside the core is mostly hydrogen, so hydrogen is fused once the temperature and density are high enough. This shell fusion (so-called because it occurs in a shell of dense material surrounding the core) causes something strange to happen to the outer parts of the star. The radiation pressure from shell fusion pushes the outer parts of the star farther from the core, making the star expand. At the same time, this material is getting less dense and farther from the heating source, so it cools. The surface of the star does exactly the opposite of the core. While the core contracts and heats up, the outer parts expand and cool down.

Helium Flash

In the lower-mass stars (less than 2–3 solar masses) the core collapses, but the pressure does not increase and the temperature does. This occurs because the gas becomes degenerate. This means that there is no longer a gas pressure associated with the material in the core, so as the temperature rises, the pressure does not. In this case, the rising temperature of the core causes the helium to be fused very quickly, creating what is known as a helium flash.

A helium flash is not a brightening of the star, however, because this occurs deep in the core of an expanding star. The energy released during the helium flash is not manifested as radiation that escapes the star, but rather, works to stabilize the core as radiation pressure, making the core no longer degenerate. Once the core has been stabilized, it begins to function like a normal helium fusing core.

Electron Degeneracy

Electron degeneracy is the reason that the cores of some low-mass stars begin to fuse helium in such an explosive way. This phenomenon is due to the fact that the core is made of up ionized gases and free electrons. As the core collapses, the gas becomes so compressed that it becomes degenerate. The free electrons are pushed closely together, and a property called the Pauli Exclusion Principle keeps them apart.

Electron degeneracy pressure occurs because no two electrons can have the same properties at the same time. (This is the Pauli Exclusion Principle: no two electrons can occupy the same quantum state at the same time.) Electrons are structureless particles, unlike protons and neutrons. Like neutrons and protons, in addition to their charge, electrons have properties like spin and angular momentum. The sheer number of electrons in the cores of these 2–3 solar mass stars creates electron degeneracy pressure. This pressure is constant and does not change with temperature (unlike gas pressure), so as the core heats up, this pressure does not increase. As a result, helium fusion begins and increases quickly.

In a normal (not degenerate) core, the increased temperature will increase fusion rates, which will slow the contraction of the core as gas and radiation pressure build up. In degenerate cores, the stopgap of increased gas pressure does not occur, so fusion rates are not regulated and slowed by a small increase in radiation pressure. Instead, fusion rates continue to increase in a runaway reaction situation until enough radiation pressure is generated to push the core outward against gravitational collapse and overcome the degenerate state. Therefore, in degenerate cores, helium fusion does not start gradually, but explosively.

The reason this occurs in lower-mass stars and not in higher-mass stars is because in higher-mass stars the core is so big that the innermost part of it can reach the temperature and density requirements for helium fusion to occur without compressing the gas to degeneracy. This means that in the cores of more massive low-mass stars (greater than 2–3 solar masses) helium fusion begins gradually in the centermost part of the core and radiation pressure builds up slowly so that the whole core is never compressed to the point of degeneracy. Conversely, smaller cores cannot reach the temperature and density requirements for helium fusion in their innermost parts. The entire core of a lower-mass star must be compressed to degeneracy just to reach the temperature requirement for helium fusion.

Helium Fusion: The Red Giant Phase

No matter the path, eventually stars that will fuse helium become red giants. Once helium fusion has taken hold and stabilized the core of the star, the outer parts of the star will begin to change. On the H-R Diagram, the star has been traveling up and to the right along the red giant branch, increasing in luminosity and size, while decreasing in surface temperature. When helium fusion has stabilized the core, the star has reached the top of the red giant branch. As the star continues its existence as a helium-fusing star, it migrates back down and to the left on H-R Diagram along the horizontal branch. This means the star is decreasing in size and luminosity as it increases in temperature.

The helium fusion process is similar to the hydrogen fusion process, in that it has many steps to it, produces new, heavier elements, and produces energy. The helium fusion process is also known as the **triple alpha process.** This name comes from the fact that helium atoms are also known as alpha particles and each cycle includes three helium atoms, or three alpha particles.

The first step of the process is two helium-4 nuclei combine to form the unstable isotope of beryllium called beryllium-8. This stage of the nuclear reaction can be written as follows:

$$He + He \rightarrow Be$$

where *He* is for helium-4 and *Be* is for beryllium-8.

The second step must occur very rapidly after the first step because beryllium-8 is a very unstable isotope of beryllium that quickly decays back into two helium-4 nuclei. In this step, the beryllium-8 atom fuses with another helium-4 atom to make a carbon-12 atom and releasing a gamma ray photon. This stage of the nuclear reaction can be written as follows:

$$Be + He \rightarrow C + \gamma$$

where *Be* is beryllium-8, *He* is helium-4, *C* is carbon-12, and γ is a gamma ray photon.

If the core is dense enough, a third step can occur. In this step carbon-12 combines with another helium-4 atom to form oxygen-16, releasing more energy. This stage of the nuclear reaction can be written as follows:

$$C + He \rightarrow O + \gamma$$

where *C* is carbon-12, *He* is helium-4, *O* is oxygen-16, and γ is a gamma ray photon.

Helium fusion, therefore, produces both oxygen and carbon in the core. When the helium in the central part of the core is reduced to a small enough fraction of the core, helium fusion will cease and the core will undergo another transition, as will the outer layers of the star.

When Helium Fusion Ends

Again, no matter what the next phase will be, red giants that deplete their helium cores do the same thing: they contract their cores further. Helium and hydrogen fusion will continue in the outermost layers of the star, with hydrogen fusion occurring furthest from the core. Ignition of these outer layers will occur during the collapse of the core. Shell hydrogen and shell helium fusion will have the same affect that shell hydrogen fusion did. The star will expand due to the radiation pressure source being brought closer to the outer layers, the increase in radiation pressure, will cause the outer layers to expand and cool and the star will move up and to the right on the H-R Diagram along the asymptotic giant branch.

This time, for all stars with masses less than about eight solar masses, the core will contract until it is degenerate. This time, the degeneracy of the core is what ignites the shell fusion layers. Eventually the expansion of the outer layers of the star begins to drag the hydrogen fusion shell with it and this shell becomes dormant. Next, the helium shell becomes dormant, then the core turns off. Following this, the gas pressure in the shells decreases so that they collapse and reignite. First, the hydrogen shell ignites and fuses hydrogen into helium. Then, a degenerate helium shell ignites in a flash not unlike the helium flash of the stars with masses less than 2–3 solar masses. This flash pushes out the hydrogen shell, causing it to go dormant again. This process repeats itself many times during the asymptotic branch life of the red giant. Each helium shell flash causes some of the outermost layers of the stars to be expelled or shed from the star. Eventually, the star is just a degenerate core with several expanding shells of material surrounding it. This is called a **planetary nebula.**

The Planetary Nebula Phase

A planetary nebula is the fate of all stars with masses less than about eight solar masses. The name planetary nebula is perhaps a little misleading. Originally astronomers of the 19th century who discovered these objects speculated that they may be solar systems forming. These astronomers had little to go on but appearance at the time. (Perhaps the most famous planetary nebula is the Ring Nebula, located in the constellation Lyra.) Now we know much more about what these objects are and how they form than the astronomers who named them. Nonetheless, the name has not been changed to reflect better the nature of these objects.

A planetary nebula is the phase reached by a lower-mass star (between about 0.4 and 8 solar masses) when it reaches the top of the asymptotic branch. The appearance of a planetary nebula is an expanding ring or shell of hot gas. Most of the radiation from the gas is from ionized hydrogen, but there is also evidence for heavier elements (such as helium, carbon, and oxygen, notably). The expansion rates of these gases are typically about 10 km/s to 30 km/s. No planetary nebulae have been observed that are significantly older than about 50,000 years. It is thought that after such a long time, the gases have cooled sufficiently that they do not radiate enough to be observed.

At the center of every planetary nebula is the cooling, degenerate core of the original star. The core is roughly 40 percent or less of the mass of the original star. The rest of the mass of the original star was ejected back into space during the planetary nebula formation.

The White Dwarf Phase

The hot degenerate core of the star will remain. This object is known as a white dwarf. The name comes from its color (white), which is due to its very high temperature, and its size. White dwarf stars are small (about the size of Earth) and very hot (from several hundred thousand degrees Kelvin down to about 5,000 degrees Kelvin). They start off at the higher temperature and, since they generate no energy within them, they cool off over time to temperatures cooler than our Sun.

These objects are called "stars" but they have no method of energy production like the other kinds of stars we've discussed. White dwarf stars produce no energy. They are hot because they used to be fusion reactors. Now they cannot even perform thermonuclear fusion because they are made of carbon and oxygen atoms and are not hot enough to fuse these atoms to produce energy. White dwarf stars will slowly cool over time.

Contrary to what might be expected, white dwarf stars do not change size as they cool. In fact, they crystallize as they cool with the carbon and oxygen atoms forming a lattice structure ("like a diamond in the sky"). The star just cools and as it does so, fades away. With no change in size, the decrease in temperature must decrease the luminosity. White dwarfs eventually become **black dwarfs** (theoretical stars that emit no light).

HIGH-MASS STELLAR EVOLUTION

In the previous section of this chapter, the evolution of stars with masses less than eight times the mass of our Sun was described in great detail. In this section of this chapter, the evolution of more massive stars will be discussed. In the main sequence phase of the evolution of a massive star, things are already different. The mass of the star is so great that it takes more outward radiation pressure to keep these massive stars from collapsing on themselves. So, the processes of thermonuclear fusion that power these stars are different.

Massive stars use the CNO cycle in addition to the process of thermonuclear fusion of hydrogen described in the last section of this chapter. The CNO cycle is a process whereby hydrogen is fused into helium using carbon, nitrogen, and oxygen nuclei, which make up about 2 percent of the material in a star (by number of atoms). Essentially, four hydrogen nuclei are fused into one helium nucleus. This fusion of hydrogen in the CNO cycle occurs in six steps where unstable isotopes of carbon, nitrogen, and oxygen are formed. Because carbon works as a catalyst in this process, the hydrogen is used up and helium is produced much faster than in the kind of thermonuclear fusion that occurs in lower-mass main sequence stars. The CNO cycle requires a higher core temperature than the hydrogen fusion process that goes on inside the cores of less massive stars. It also requires a higher pressure and the presence of at least a critical amount of carbon atoms.

Stars with masses greater than eight solar masses have hotter cores and greater pressures within those cores. The faster rate of fusion means these stars will use up their hydrogen fuel faster than lower-mass stars. So, more massive stars spend less time on the main sequence than less massive stars.

Helium Fusion: The Red Giant Phase

Just like lower-mass stars, massive stars transition to a helium fusion phase after exhausting the hydrogen in their cores. This phase begins with hydrogen shell fusion turning on, which results in the expansion of the outer shell material of the star, making the star a red giant. In the case of the most massive stars, the size of the giant star is considerably larger than for lower-mass stars. Astronomers call these most massive helium fusing objects red supergiants.

Again, similar to the main sequence phase for massive stars, the red giant phase is short-lived. Helium core fusion begins when the core collapses further and reaches the critical temperature and density necessary for helium fusion to begin. Just like hydrogen fusion begins in all stars, massive stars begin helium fusion gradually, starting in the core where the temperature and density are high enough first and moving outwards until the entire core is actively fusing helium into carbon.

This phase of a massive star's existence will last only a few hundred thousand years, whereas this phase of lower-mass stars lasted millions of years. Following the core becoming inert carbon, shell hydrogen and shell helium fusion will begin outside the core, again causing the star to increase in size. The core will undergo a small collapse causing it to heat up and become more dense.

Carbon Fusion

The result of this second collapse of the core of the massive star is carbon fusion. Carbon fusion requires a core temperature of 600 million K. Carbon fusion can be through helium capture (carbon fuses with helium to create oxygen) or with other heavy elements (carbon fuses with oxygen to form silicon). Carbon fusion will provide the necessary energy source to put the star back into gravitational equilibrium until carbon fusion ceases, after a few hundred years.

Following the carbon fusion phase, hydrogen, helium, and carbon shell fusion will begin in the outer layers of the core and the core will collapse again, igniting the oxygen ash and beginning the death sequence for the massive star.

Heavy Element Fusion

Fusion of heavy elements into heavier elements occurs for ever shorter and shorter periods of time. These fusion processes are complex and can occur simultaneously. The pattern of how these phases begin and end is the same, however. Fusion occurs until the element being fused runs out, then fusion stops. The outer shell layers of the core ignite when the core ceases fusion, the core collapses, heats up, and begins to fuse the ash left behind from the previous reaction.

This can continue until the core becomes iron ash. Fusing iron nuclei requires energy, so the star cannot create a new energy source by fusing iron. Nor can the star produce energy by fission of the iron nuclei, as that also requires an energy input. As a result, the star will not be able to return to gravitational equilibrium. Instead, the star will collapse on itself in a catastrophic explosion called a supernova.

SUPERNOVAE

When a massive star runs out of elements that it can fuse in its core, it is no longer in balance. While the star is producing energy in its core, it can balance, with an outward force, the force of the weight of the star so that the star does not collapse. When the core of a massive star reaches the point that it contains insufficient materials for fusion processes that produce energy, the star collapses.

∙∙∙

Supernovae

The name "supernova" comes from the fact that early astronomers who watched the skies for signs from the heavens called anything that appeared in the sky that was not known before a "nova," which is Latin for "new." The first supernova recorded by humans was recorded by Chinese and Native Americans. Based on the expansion rate of the supernova remnant, the supernova recorded is thought to be that which created the Crab Nebula. The only other supernova that occurred in our galaxy, the Milky Way, known to be observed by humans is one observed by Tycho Brahe in 1632. The word supernova has its origin in the early 20th century, according to the *Oxford English Dictionary*, however.

∙∙∙

The outer layers of the star fall, unsupported, and crash into the iron core. Depending on the size (or mass) of that iron core, one of two things will happen. The core may collapse, but remain supported against further collapse by the force of repulsion between neutrons (neutron degeneracy), leaving a remnant called a neutron star. Or, the core may be too massive to remain supported by the strongest force in the universe and will collapse into a black hole.

SUPERNOVA REMNANTS

When a massive star collapses in a supernova, the material of the star bounces off of the inert iron core and spreads outwards from the remnant neutron star or black hole for thousands of years after the explosive end of the massive star's existence.

The Expanding Shell

The material is heated to extremely high temperatures and even more massive elements are formed in the process, using some of the energy of the supernova event. The hot, low-density gas that was once the bulk of a massive star expands into the near-vacuum of space interacting with whatever dust or gas is in its path. The expanding shell of extremely hot, low-density gas can be observed thousands to millions of years after the supernova event.

Such supernova remnants are often mistaken for other types of objects, but fall into the category of "nebulae." The most well-known supernova remnants are several hundreds to several thousands of years old like the Crab Nebula, the Veil Nebula, and the Cygnus Loop Nebula.

The Remnant Core

The two types of supernova remnant cores that can exist are neutron stars and black holes. Neutron stars are not really stars, per se, as they don't produce any energy in their cores. They are merely hot bodies that radiate their heat and cool over time. A neutron star, similar to a white dwarf, is made of inert material that is supported from collapse by the fact that the material is also degenerate.

Whereas a white dwarf is supported by electron degeneracy pressure, a neutron star is supported by neutron degeneracy pressure. Electron degeneracy pressure is a pressure or force due to the quantum rule that no two electrons can occupy the same quantum space at the same time. That means that no two electrons can have the same position, charge, spin, momentum, and magnetic moment at the same time.

A good analogy of this "pressure" is to imagine how two people wearing the same costume might behave if they appeared at the same costume party. They would likely stay as far apart as possible at all times. This is how electron and neutron degeneracy pressure work to support these types of stars against collapse. In white dwarf stars, the pressure is due to electron properties, whereas, in neutron stars, the pressure is due to neutron properties. Neutrons have fewer quantum states (fewer possible properties in common) than electrons, so neutron degeneracy pressure is stronger than electron degeneracy pressure.

In fact, the force that governs neutron degeneracy pressure is so poorly understood that astronomers can only estimate an approximate upper limit for the mass of a neutron star. The best models suggest that a neutron star with a mass greater than three times the mass of our Sun would not be able to resist gravitational collapse and would become a black hole. Experimental physicists have not been able to reproduce or simulate scenarios that mimic the conditions inside neutron stars. The pressures inside neutron stars are 10–100 times greater than anything that can be created in a laboratory on Earth.

One type of neutron star found is a pulsar. This type of object is unique in its electromagnetic signature. Pulsars emit radio radiation in regular, pulse-like, beats. The regularity of the emission of a pulsar is incredibly near to exact.

Pulsars are neutron stars with strong magnetic fields that are not aligned with their rotational axes. Because of this misalignment, material, guided by the magnetic field associated with the object, falls on a spot on the surface of the star causing it to heat up to a temperature higher than the rest of the star. Since this spot is not located at the axis of rotation, the hotspot rotates with the star. The hotspot radiates more electromagnetic radiation than the rest of the star. The observer sees something that has the appearance of a spinning flashlight with the lit end rotating into and out of view.

The rotation rate of a pulsar is inversely proportional to the age of the pulsar. Just after the supernova event the star is spun up to the highest rate at which it will spin. Over time, it slows down and the rate at which it slows down can be used to determine how long it has been since the supernova event that created the pulsar.

Black holes are objects that are so dense that at their surfaces (called event horizons), the escape velocity is greater than the speed of light. This means that although these objects are collapsed cores of massive stars, they radiate no electromagnetic energy. Since no light escapes, the name "black hole" was given to these objects.

Black Holes

There is a common misconception that black holes are extremely massive objects, but actually, it is not that they are massive, but that they are compact that they have such interesting properties. It is the amount of mass in such a small space that makes them unique objects. For the Sun to become a black hole, for example, it must shrink in size so that all its mass fit in a volume of radius 3 km. (For comparison the radius of the Sun is currently 700,000 km.) Since the mass of the Sun does not change in this scenario, the gravitational pull due to the Sun does not change either. So, if the Sun were to become a black hole (which it cannot, since it is not massive enough to end up with an iron core), the solar system would *not* be sucked into it.

Black holes do not radiate any electromagnetic energy, so how can astronomers tell they are there? While black holes don't radiate electromagnetic energy, they do have a gravitational field associated with them. This gravitational field affects the way objects near the black hole will move. The more massive the black hole the faster objects near it will move.

Also, some black holes attract material that is accreted onto the black hole by way of an accretion disk of material. This disk of material gets heated to extremely high temperatures and can radiate electromagnetic radiation (since it is not part of the black hole). Sometimes, astronomers can identify black holes by identifying an accretion disk.

Interstellar Medium Feedback Processes

The **interstellar medium** is an important part of the cosmos. It is comprised of both molecular and atomic gas as well as dust. The interstellar medium is literally the stuff between the stars. Space is not entirely empty, although the density of the material in space is so low one would certainly not be completely wrong to think so. Most of the interstellar medium is in the form of atomic hydrogen gas. The next greatest constituent is molecular hydrogen gas, then other gases, then dust. The dust is mostly carbon based; however, there is some evidence of silicon-based dust in the universe (particularly in our Galaxy). More detail about the interstellar medium can be found in chapter 5.

Stars form and evolve and eventually cease to exist. Stars cannot form without the necessary ingredients. These ingredients exist in galaxies and also between galaxies, but if stars were not recycled to a certain extent, the universe would run out of ingredients very quickly. Perhaps we would already have used up all the available ingredients in the universe. Luckily, however, the universe naturally recycles a large fraction of the material tied up in stars. And the best recyclers are the most massive stars, which return almost 90 percent of the materials they use during their existences to the interstellar medium in the form of a supernova remnant.

The material expelled by stars at the ends of their lives (in the form of planetary nebulae and supernova remnants) becomes part of the interstellar medium (the material between the stars). The interstellar medium consists of atomic and molecular clouds of gas, and dust. These are the ingredients necessary for the formation of stars.

The material contained in the interstellar medium is enriched through many different processes associated with the formation and evolution of stars. Some material is ejected from stars early in their existence, while some comes only at the end of a star's life. Each stage where stars contribute to the composition of the interstellar medium is important. In this section, we will explore several of the most well-understood processes and describe how they contribute to the composition of the interstellar medium.

T Tauri Phase

Low-mass stars (like the Sun) lose mass both at the beginning and end of their existence. During the gravitational collapse of a cloud of interstellar medium during the formation of low-mass stars, the collapsing star is both accreting mass from the collapsing cloud and ejecting mass into the collapsing cloud. During its formation, the protostar spins faster as it collapses and the material of the cloud forms a disk around the protostar. Material from the disk accretes onto the surface of the star, and the star begins to expel material in the form of jets that form at the poles along the axis of rotation of the protostar.

The first such star observed was in the constellation Taurus. Until it was studied well, it was not understood what the physical mechanism was that caused its unusual properties of irregular variable brightness and emission lines. Now, astronomers understand not only that there are many stars that have the same unusual properties, but that all stars like our Sun go through a phase in their evolution that causes them to exhibit these properties. The material expelled by the jets of a star in the T Tauri phase includes both gas and dust.

Protoplanetary Nebulae

Otherwise known a proplyds, protoplanetary nebulae occur around some T Tauri stars. These nebulae are thought to be the birth places of solar systems. They are dense and dusty, so that they are apparent in the visible part of the electromagnetic spectrum only by the light they block, but are very bright in the infrared, indicating that they are much cooler in temperature than the star. The protoplanetary nebula is, itself, a recycling center, taking materials that the protostar is ejecting or is not accreting, and forming planets and other bodies that will be part of a system of material around the protostar. It is primarily the T Tauri part of the proplyd that is ejecting material into the interstellar medium by way of its jets. The material expelled by a proplyd includes both dust and gas.

Novae

When gas falls on a low-mass white dwarf but does *not* increase the mass of the white dwarf to the point that it exceeds the Chandrasekhar Limit (1.4 times the mass of the Sun), the white dwarf undergoes a nova. A nova is a brightening of a white dwarf due to fusion of hydrogen or helium occurring on the surface of the star.

To create a nova event, material must be accreted onto the star. That material will lie in a thin layer on the surface of the star. When the accreted material reaches a critical density, fusion occurs simultaneously all over the star throughout the entire layer of accreted material and the star appears to brighten briefly (for several minutes to several days) until all fusible material has been exhausted.

When all the fusible material has been fused, the star returns to its stable state of inert material supported by electron degeneracy pressure. This period of surface fusion results in a massive outflow of fusible material because there are no outer layers of a star to press down and resist the radiation pressure created by the epoch of massive fusion. The hot gas that escapes the surface of the star becomes part of the interstellar medium again.

This process can repeat and continue to repeat until the mass accreted by the white dwarf star causes its total mass to exceed a mass equivalent to 1.4 times the mass of our Sun. This mass of 1.4 times the mass of the Sun is referred to as the Chandrasekhar Limit. If the mass of a white dwarf star exceeds the Chandrasekhar Limit, the force of electron degeneracy pressure will fail to prevent the star from total collapse. The result of this scenario is a type of supernova that leaves behind no remnant.

Supernovae

As discussed earlier in this chapter, supernovae are another method of recycling the materials in the universe. At the end of a star's life, a large fraction of the material that was part of the star becomes part of the interstellar medium and can become part of a new star or stars. There are two main types of supernovae that have been observed. These are known as Type I and Type II.

Type I

A Type I supernova is a supernova whose spectrum contains no hydrogen emission lines. Additionally, the light from this type of supernova decays rapidly. That is to say that the initial brightening from the explosion disappears quickly. The decay of the light from a supernova of this type, however, is also nearly constant. That is to say that it fades away at a constant rate.

Astronomers have learned that this type of supernova comes from a different initial physical scenario (progenitor) than a Type II supernova. Type I supernovae are the result of a binary star system that includes a white dwarf. Mass is transferred from the companion to the white dwarf when the companion reaches the red giant phase and fills its Roche Lobe causing its outer shell to be accreted onto the white dwarf. When the mass of the white dwarf exceeds the Chandrasekhar Limit (1.4 times the mass of the Sun), the white dwarf can no longer prevent gravitational collapse and a supernova occurs. This type of supernova is thought to be completely destructive and leaves behind no known observable remnant.

This type of supernova has such a particular type of progenitor, a white dwarf, that the peak luminosity of a Type I supernova is constant from supernova event to supernova event. For this reason, supernovae of this type are

● ●

Transfers of Mass Between Stars

When stars are physically close to one another they can transfer mass. This is because of the fact that both stars are massive enough to attract materials that are close to their surfaces. But, since gravity is an attractive force, there is a point between the two stars where the total (net) force on an object is exactly zero. This point in space where the force of each star is equal to the force of the other is called an equilibrium point because an object can stay in this spot indefinitely. If one star expands so much that it expands past this equilibrium point, the material will be attracted to the other star, so it will be accreted onto its surface.

● ●

considered standard candles and are used to measure the distance to distant galaxies. Knowing how luminous the supernova is and how bright it appears, astronomers can calculate the distance to the supernova.

Type II

A Type II supernova is one whose spectrum contains prominent hydrogen absorption lines. Another aspect of a Type II supernova is how the brightness decays. A Type II supernova decreases in brightness quickly at first, then stabilizes for about 75 days, then drops off significantly faster. By the spectrum and light curve of a supernova, one can identify the type.

Astronomers also now know that the progenitor of a Type II supernova is different from that of a Type I supernova. A Type II supernova is the end of the existence of a star with mass greater than nine solar masses. Because all stars with a mass greater than nine solar masses will supernova, and since stars can have masses up to 100 solar masses, the peak brightness of a Type II supernova is not the same for every supernova event.

The Circle of Stellar Life

The material that makes up the universe is recycled constantly through the formation and evolution of stars. Clouds of atomic gas cool and condense, forming cold, dense clouds of molecular gas. Within these clouds of cold gas, stars can form. It takes just some push or pull or magnetic field to cause the cloud to fall into gravitational collapse. Once the cloud is collapsing, the road to becoming a star (or many stars) is an inevitable one.

The formation of a star involves many different stages. In the beginning there is just dust and gas. Slowly, over time, the dust and gas form a cloud of material containing all the right ingredients for star formation. Then, something happens to cause the cloud to gravitationally collapse. The cause can be a nearby supernova, or stellar winds from a massive star going through an early stage of mass-loss, or it could be the galaxy's magnetic field getting twisted up and causing the cloud to collapse. Whatever the reason, the cloud begins to collapse and stars begin to form.

The first objects formed are protostars. These are massive objects that glow because they are hot. Protostars are hot because the cloud is condensing and the material that made up the cloud is getting hotter. The energy of the gravitational collapse is converted into thermal radiation. Protostars are most easily seen in the infrared because their temperatures, compared to fully formed stars, are cool.

A protostar will collapse until something stops it. Gravity is working to pull the material of the protostar together. When the core of the star becomes dense enough and hot enough to undergo thermonuclear fusion, the added outward force of radiation pressure will act to counter the gravitational collapse. At this point the object becomes a main sequence star.

Depending on the mass of the star, it may or may not go through a T Tauri phase where material is spewed back into the interstellar medium through jets of material that form at the rotational axis poles of the star. More massive stars tend to have strong stellar winds that cause a good fraction of the material that would be the outer layers of the star to be expelled back to the interstellar medium.

Once fully stabilized, the star will simply follow its evolutionary sequence, and the interstellar material that was used to make the star is almost fully recycled back into the interstellar medium. A low-mass star (like our Sun) will end up as a planetary nebula and white dwarf. A good part of the material of the star will be expelled back into the interstellar medium in the form of a planetary nebula, with only a fraction of the mass of the star tied up in a white dwarf. A high-mass star will end up as a supernova with a neutron star or black hole remnant. Again, the bulk of the material of the massive star will go back to being part of the interstellar medium as a supernova remnant with a small fraction of the mass of the original main sequence star (progenitor) tied up in the remnant neutron star or black hole.

The material from the star that goes back into the interstellar medium will be used in the formation of new stars later. In fact, the event that marks the end of one star's existence (a supernova) can, ironically, mark the beginning of another star's existence by triggering the collapse of a nearby cloud of interstellar material.

RECOMMENDED READINGS

Asimov, Isaac. *Asimov on Astronomy.* New York: Bonanza Books, 1988.

Bennett, Jeffrey D., Megan Donahue, Nicholas Schneider, and Mark Voit. *The Cosmic Perspective.* 5th ed. San Francisco: Benjamin Cummings, 2007.

DeGrasse Tyson, Neil, Charles Tsun-Chu Liu, and Robert Irion. *One Universe: At Home in the Cosmos.* Washington, DC: Joseph Henry Press, 1999.

Freedman, Roger, and William J. Kaufmann III. *Universe.* 8th ed. New York: W.H. Freeman Company, 2008.

Hawking, Stephen. *On the Shoulders of Giants.* Philadelphia: Running Press, 2002.

Jastrow, Robert. *Red Giants and White Dwarfs.* New York: W.W. Norton and Company, 1990.

Seeds, Michael A. *Astronomy: The Solar System and Beyond.* 5th ed. Pacific Grove, CA: Brooks Cole, 2006.

4

Variable Stars and Multiple Star Systems

There is an entire class of stars that do not fit the working definition of stars we've used throughout this volume. These stars are not in equilibrium. These stars are in an extended period of transition. These stars are called "variable stars." They are important for many reasons. Some of them simply tell us about the physical nature of variability in stars, while others can be used as standard candles and can be used to determine the distance to objects in the Galaxy or universe.

There are two basic types of variable stars: intrinsic and extrinsic. Intrinsic variable stars are divided into four groups: pulsating, eruptive, cataclysmic, and x-ray. Extrinsic variable stars are either eclipsing or rotating. Each of these types of variable stars will be discussed in this chapter. Details about how, if at all, the variable property of the star fits into the evolutionary sequence will also be discussed.

INTRINSIC VARIABLE STARS

These are stars whose brightness changes due to the star itself changing in luminosity. An intrinsic variable star changes in luminosity because it expands or contracts or because some internal process changes the energy output of the star. In the category of intrinsic variable stars we include stars that pulsate, stars that erupt, and stars that have outbursts on their surfaces or accretion disks. This includes supernovae and novae (discussed in the previous chapter) as well as RR Lyrae and Cepheid variable stars, which are used as distance indicators.

Pulsating Variable Stars

A pulsating variable star is a star that is undergoing a periodic expansion and contraction of its outer layers. This change in the shape of the star changes its energy output and surface temperature. The pulsation of such a variable star can be radial (occurring symmetrically over the entire surface of the star) or not. Most of the types of pulsating variable stars are radially pulsating. These stars are variable because they are in a stage of their evolution that causes the star to undergo gravitational instability and radial pulsation.

The Instability Strip

Stars become unstable when they make the transition from fusing one kind of element in their cores to another. The most well-understood transition is the transition from main sequence star to red giant, which signifies the transition from core hydrogen fusion to shell hydrogen fusion and core helium fusion. All stars greater than about 0.4 solar masses go through this transition. On the H-R Diagram, this transition region is called the instability strip.

The lower mass stars pass through this instability strip at a slower pace than the higher mass stars. The lower the mass of the star, the longer the stages of evolution. Inside the star, during this transition, the core of the star ceases to fuse hydrogen and begins to gravitationally collapse. As it does, the outer parts of the core get dense enough and hot enough to fuse hydrogen and a process known as shell hydrogen fusion begins.

Shell hydrogen fusion causes the outer layers of the star to expand into a red giant. Shell hydrogen fusion can also stop, as the thin shell of hydrogen is quickly depleted, then re-start as more of the core material collapses, drawing more of the shell material in closer to the core where it can fuse. Core helium fusion starts up similarly. It can start, then stop and start again.

When a star transitions to helium fusion in its core in this start-stop-start manner, it goes in and out of stability, causing it to shrink and expand. Since the material of the star does not escape it, the pulsations are regular as the mass of the core slowly increases until it reaches a stable equilibrium point where fusion can occur continuously, using up the fusible material in the core. This means that some stars will become variable stars for part of their existence as a kind of phase of their evolution. So, variable stars are not variable all the time.

RR Lyrae Stars

One type of intrinsic pulsating variable star that is well-studied and understood is the RR Lyrae variable star. The first star of this type was found in the constellation Lyra. The star designated RR Lyrae is the first star to be observed to have these properties. RR Lyrae stars are stars of spectral types A–F that are on the lower portion of the instability strip found between the main sequence and the red giant branch for these types of stars.

An RR Lyrae star has a very low **metallicity** (meaning a very small amount of elements more massive than helium) and is usually found in a globular cluster or other location typically populated by Population II stars. RR Lyrae stars have a characteristic period of variation of that is usually 0.5 days with a range of about 0.1 to 2 days. The amount of variation for these stars is usually 1 magnitude (with a range of about 0.3 to 2 magnitudes).

Because RR Lyrae stars are in such a specific part of the H-R Diagram, the average absolute magnitudes of these types of stars is known. Therefore, measuring a RR Lyrae's average apparent magnitude allows one to calculate the distance to an RR Lyrae star. This type of measurement (where the luminosity or absolute magnitude is known) is called a "standard candle measurement." It is as if you were looking at a series of lamps located at different distances from you. Each one has the same type of lightbulb in it. So, simply observing which lamps appear brighter tells you which ones are closer. Knowing exactly how much light you are receiving and exactly how much light is emitted from each lamp, one can calculate the distance to the lamp.

Population II Stars

Population II stars are stars that occupy the spheroidal components of our Galaxy (i.e., the **bulge** and the **halo**). Due to their low content of elements heavier than helium (low metallicity), they are thought to be stars that formed longer ago than stars that contain more elements heavier than helium. Since all elements heavier than helium were formed in stars either through fusion in their cores or through fusion processes that occurred during their cataclysmic ends, the presence of elements heavier than helium indicates that these stars formed in an interstellar medium enriched by the recycling of materials through the evolution of a previous generation of stars. So, low metallicity stars are necessarily older than high metallicity stars. The implications of this information on how the Galaxy must have formed are discussed in detail in chapter 6.

Cepheid Variable Stars

Another type of intrinsic pulsating variable star is the Cepheid variable star. These stars are known to be yellow giant or supergiant stars. On the H-R Diagram they are found in the upper portion of the instability strip. These stars are subdivided into two groups Type I and Type II.

The Type I Cepheid variable stars are also known a Delta Cephei stars. The first object discovered to exhibit these properties was Delta Cephei, the fourth brightest star in the constellation Cepheus. These stars are Population I stars found in the spiral arms or disk region of the Galaxy. Delta Cephei stars have periods of variation that range from about 5 to 10 days. Delta Cephei stars are found on the upper right part of the instability strip, as they are more luminous than the Type II Cepheid variable stars.

· ·

Population I Stars

Population I stars are stars with high metallicity. That is, they contain elements in their spectra that indicate that they were formed from an interstellar medium that had been enriched with elements heavier than helium. In other words, Population I stars are formed from recently recycled material. Therefore, they are necessarily younger than Population II stars.

· ·

The Type II Cepheid variable stars are also known as W Virginis variable stars. The first object to be discovered to exhibit these properties was the star in the constellation Virgo named W Virginis. The Type II Cepheid variable stars are distinctly different from the Type I stars because they are Population II stars found in the spheroidal parts of the Galaxy. W Virginis stars have periods of variation that range from about 10 to 30 days.

Both of these types of Cepheid variable stars were, and still are, used to measure distances. Cepheid variable stars have the unique property that their period of variation in luminosity is related to their actual average luminosity. That is, the longer the period of variation, the brighter the star. For this reason, they are used as (varying) standard candles. To get the distance to a globular cluster or galaxy containing a Cepheid variable, measure its period of variation and its average apparent magnitude. Knowing the period allows one to determine the average luminosity, or average absolute magnitude using the period-luminosity law. Knowing both apparent and absolute magnitude, the distance can be calculated using the distance modulus formula:

$$m - M = 5 \log D - 5$$

where m is apparent magnitude, M is absolute magnitude, and D is distance in parsecs.

Mira-type Variable Stars

Another well-studied type of intrinsic pulsating variable star is the Mira-type variable star. These stars are giant stars. They can vary in brightness by more than 2.5 magnitudes and their period of variation is from about 80

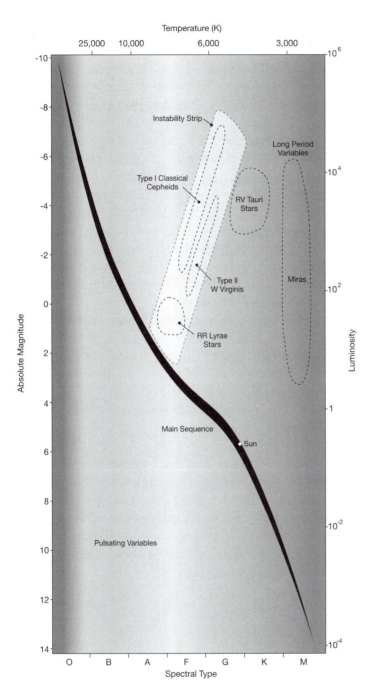

Figure 4.1 The above diagram is an H-R Diagram showing the instability strip. The different regions of the instability strip that correspond to particular types of variable stars are labeled. The above H-R Diagram shows where many variable star stages can be found. [Jeff Dixon]

to 1,000 days. Mira-type variable stars are known as long-period variable stars. The first object discovered to have these properties was a star named Omicron Ceti, the 15th brightest star in the constellation Cetus. Later, it was named "Mira" (meaning "the wonderful"). On the H-R Diagram, Mira-type variable stars are found at the high luminosity end of the asymptotic giant branch.

ZZ Ceti Variable Stars

The ZZ Ceti star is the only type of non-radial pulsating intrinsically variable star discussed in this volume. A ZZ Ceti star is a white dwarf star that changes in brightness only slightly (0.001 to 0.2 magnitudes) over very short periods of time (30 seconds to 25 minutes). The first star discovered to exhibit this type of behavior was the star named ZZ in the constellation Cetus.

There are three different types of this kind of variable star, distinguishable only by their spectra. The three types reflect the temperature and composition of the white dwarf star. The first is DA for a white dwarf star with a hydrogen-rich outer layer. A DA star is distinguished by strong hydrogen absorption lines in the spectrum. The next is DB for a white dwarf star with a helium-rich outer layer. A DB star is distinguished by strong neutral helium absorption lines. Finally, a DO white dwarf star is one with a helium-rich outer layer that is distinguished by ionized helium absorption line, indicating that the star is much hotter than the DB type star.

The short period and slight changes in brightness, as well as the fact that usually several periods of variation are observed at once indicate that these ZZ Ceti stars are not radially pulsating. Rather, these stars are experiencing non-radial pulsations that are warping the surface of the star asymmetrically.

Eruptive Variable Stars

Eruptive or cataclysmic variable stars are stars that have occasional violent outbursts caused by activity in its chromosphere or corona or enhanced stellar wind or coronal mass ejections. A nova and a supernova would be considered types of cataclysmic variables. Most of the types of eruptive variable stars discussed in this chapter are repeating variable stars, unlike supernovae, but they are also mostly irregular variables as their variation in luminosity is usually due to a physical process that is irregular.

T Tauri Variable Stars

T Tauri stars were discussed briefly in the last chapter in the context of recycling matter in the universe. This type of variable star is an irregular variable star, which means that there is not really a period of variation range or brightness variation range. These stars are identified by their spectra. They have the spectra of low-temperature (G–M spectral class) stars with strong emission and absorption lines. In particular, lithium is observed in the spectra of these stars, indicating that they are very young objects (probably less than 10 million years old).

Another indication of the youth of these objects, is the fact that they are enshrouded in an envelope of dust and gas and have large protoplanetary accretion disks. The luminosity of a T Tauri star changes, most likely, due to instabilities in the disk, violent activity in the atmosphere of the star or moving clouds of dust and gas. T Tauri stars are an early phase in the evolution of low-mass stars, like our Sun.

UV Ceti Variable Stars

UV Ceti stars are red or orange low-mass dwarf stars. These stars exhibit rapid, irregular variations in brightness that are not predictable. UV Ceti stars are also known as flare stars. Their luminosities can change abruptly in many different parts of the electromagnetic spectrum at once or in sequence. The spectra of UV Ceti variable stars include hydrogen and calcium lines, indicating activity in the chromospheres of these stars.

Astronomers equate the activity in UV Ceti to the activity we observe on our own Sun. On the Sun, flares are caused by magnetic energy being released through a rapid heating of plasma, which manifests itself as a spike in brightness of a portion of the solar material. In UV Ceti variable stars, the flares must be significantly larger (compared to the star from which the flare originates, since most UV Ceti stars are cool K–M spectral type dwarf stars) and tend to cause spikes in the ultraviolet-blue portion of the electromagnetic spectrum.

R Coronae Borealis Variable Stars

This type of variable star is distinctive in that it varies in brightness by dimming, not brightening. The R Coronae Borealis variable stars are hydrogen-deficient and carbon-rich spectral type F or G supergiants. These stars experience intermittent fluctuating minima in their luminosities. These declines in brightness are due to carbon-rich dust clouds that obscure the photosphere of the star.

The observed minima occur randomly and unpredictably; however, each successive decline in brightness within a minimum is due to the formation of a dust cloud. As the dust cloud moves out of the way, it exposes the photosphere and the full luminosity of the star is observed again. It is thought that the carbon dust clouds form from **ejecta** (material ejected by the star). Moving outward, the material cools and condenses into a dust cloud when it reaches a distance where the temperature is low enough for carbon to condense (about 20 stellar radii).

This process is thought to be a source of interstellar dust in galaxies; however, this stage of stellar evolution is short-lived, lasting only about 1,000 years, which, in the course of the existence of a star like the Sun, is a *very* short time. It is also uncertain where this phase lies in the evolutionary sequence of a star. R Coronae Borealis stars are thought to be either the product of the merging of a binary white dwarf system or a final helium shell flash where a pre-white dwarf star inflates its remaining small envelope resulting in a "reborn" asymptotic giant branch star. Either way, however, this phase is associated with medium- to low-mass stars.

FU Orionis Variable Stars

The prototype star of this class of stars is found in the constellation Orion. Its unusual behavior that defines it as a type of variable star is that it brightened by 6 magnitudes (100 times its original brightness) over a period of about a year. During this brightening phase, a nebulous glow was observed surrounding the star. The brightening of the star was not a smooth uniform brightening, but rather it was due to multiple bursts of increasing brightness that do not diminish. The star's brightness no longer varies, and there are now several (more than 10) examples of other stars that have exhibited similar behavior.

These stars are pre-main sequence stars in early stages of their development. The observed increase in brightness is thought to be caused by thermal instabilities that arise in the inner portions of the accretion disk surrounding the star. The thermal instability causes the gas in the disk to become ionized, and releases a burst of radiation when the hot material is accreted onto the star. The amount of matter accreted determines the length and brightness of the burst, so it is not possible to predict the length or duration of the brightening.

Although it has not yet been observed, current theories suggest that this epoch of brightening from accretion of hot material is part of a repetitive process. The cycle may take much longer than a human lifetime, with long periods of relative quiescence (inactivity). Astronomers are still studying FU Orionis in order to learn more about what caused its brightening.

Wolf-Rayet Stars

Wolf-Rayet stars are a class of stars that are identified by their spectra, rather than their variability. They are variable stars, and their variation in luminosity is irregular. Wolf-Rayet stars are hot, massive stars in an advanced stage of evolution. During this stage of evolution, the massive stars are expelling material in the form of massive stellar winds. Wolf-Rayet stars are thought to be O stars that have had their outer hydrogen envelopes stripped due to their being a member of a close binary system (such systems are described in detail in the next section of this chapter). The massive helium core is exposed. Depending on the initial mass of the star, the core can include other elements.

There are two subclasses of Wolf-Rayet stars that can be identified by their spectra. The type known as WN have prominent emission lines of helium and nitrogen. The type known as WC have prominent carbon, oxygen, and helium lines. The difference between these two types is most likely due to the initial mass of the star in question. More massive ($M > 25$ $M_\odot$) stars may reach this phase with heavier elements formed since they may be fusing hydrogen using the CNO cycle, rather than the proton–proton chain, whereas less massive stars (9 $M_\odot < M < 25$ $M_\odot$) may not have as many heavy elements formed when they reach this stage in their evolution.

U Geminorum

This class of variable stars is defined by their irregular and extremely bright bursts of luminosity. These stars are dwarf stars that are in a close binary star system. The stars in this binary system are so close that mass transfer is occurring between the stars. The white dwarf star is accreting matter from its binary partner. As the material leaves the partner star, it accumulates in an accretion disk around the dwarf star. Periodically, an instability occurs in the disk (or in the other star) that causes the rate of accretion of material onto the dwarf to increase significantly. This increased rate of accretion results in an explosive outburst, which brightens the star for a short period of time.

The brightening in this case is due to the same phenomenon that causes a nova. Material accumulates on the surface of the white dwarf, causing the white dwarf to shrink and condense. When the density of the material accreted onto the star is high enough for fusion to occur, it does so all at once. This creates a very short, very bright burst in energy output from the star.

Magnetic Cataclysmic Variable

A magnetic cataclysmic variable star is a particular type of system where the primary member is a white dwarf star and the secondary star is a red dwarf star. In this type of system, however, the white dwarf has a very strong magnetic field that completely dominates the accretion of material in the system. The strength of the magnetic field of the white dwarf star can range from relatively low to relatively high. The stronger the magnetic field strength, the better synchronized the orbit of the white dwarf is to the orbital period of the system. This synchronous rotation means the same side of the white dwarf faces the red dwarf all the time.

In magnetic cataclysmic variable star scenarios, the magnetic field of the white dwarf is so strong that when the accreted material gets close enough to the white dwarf to interact with its magnetic field, the material is simply accreted along the magnetic field lines of the star, rather than forming an accretion disk. Since the material is charged and moving along a magnetic field, its path is a spiral, not linear. The result of this motion is **synchrotron radiation** in the form of x-rays. These star systems also emit polarized light due to the strong magnetic fields involved.

Symbiotic Stars

Symbiotic stars are star systems where two stars are gravitationally bound. The typical components of a symbiotic star system are a red giant and a white dwarf. The two stars are usually surrounded by nebulosity that is thought to be material from the red giant that is being lost through stellar winds or pulsation. The variability observed in these systems is caused by mass loss from the red giant to the white dwarf resulting in a nova. In this system, however, the material being lost from the red giant is not accreted onto the white dwarf through an accretion disk. In fact, the two members appear to be detached. No accretion disks have been observed in these systems. Rather, the material is thought to fall on the surface of the white dwarf.

The result of this activity is that these stars experience irregular variations in brightness. The change in brightness can be as much as 4 magnitudes. The frequency of these variations is not regular and is separated by several hundred days. This class of objects is not very homogeneous; no two objects in this category share all their characteristics.

X-ray Variable Stars

This type of variable star is also a binary system, but in this case the binary system includes a neutron star or black hole in addition to another star. These systems are characterized by the fact that they emit x-rays. The x-rays come from material in the accretion disk surrounding the compact object being heated to extremely high temperatures. If the star in the system is like our Sun, the system is classified as a low-mass x-ray binary (LMXB); if the star is greater than 10 $M_\odot$, the system is classified as a high-mass x-ray binary (HMXB). This class of star experiences irregular variations in brightness with no limit to the range of brightness changes or period length.

EXTRINSIC VARIABLE STARS

An extrinsic variable star is one whose brightness fluctuates for reasons unrelated to the physical nature of the star. There are two basic types of stars that experience such changes in brightness. There are eclipsing binary stars and rotating variable stars.

Eclipsing Variable Stars

Eclipsing variable stars are stars that are part of a binary system where one star passes in front of another, as seen from Earth. When the hotter star is eclipsed by its cooler companion, the total luminosity of the system decreases. Eclipsing variable stars have very distinctive light curves with flat-bottomed minima and maxima that are separated by very regular intervals of time. The luminosity changes vary from system to system and depend on both the surface temperatures of the stars in the system as well as the distance between the two stars. Eclipses can occur in cataclysmic binary systems as well.

Algol

Algol type stars are eclipsing binary stars that form a close binary (where mass transfer is possible), but are detached or semidetached. They have periods of 5 hours to 30 years and experience brightness variations of several magnitudes. If mass transfer takes place, it is by direct accretion, rather than accretion disk, as in symbiotic variable stars.

Beta Lyrae

This type of system is a close binary star system that is both an ellipsoidal variable and an eclipsing binary. An ellipsoidal variable star is one whose shape has been distorted due to gravitational attraction to its companion so that as it rotates, its surface area, as seen from Earth, varies, causing its brightness to vary. The variation of brightness in a Beta Lyrae type of variable is small (a few magnitudes) and smooth, though the variation is related to both the eclipsing and ellipsoidal properties of the system. While the circumstances of this type of variable seem unusual, Beta Lyrae is thought to be an example of one phase in the evolution of certain close binary systems.

W Ursae Majoris

This type of variable star is a more extreme type of binary system than the previous two. In this case, there is a binary system, but the stars in it are so close, they are almost touching. This type of binary system is called a **contact binary.** Since the stars are so close, this type of variable star is characterized by its short period and similar change in brightness (between the primary and secondary minima).

Rotating Variable Stars

A rotating variable star is one whose brightness changes as it rotates due to dark spots on its surface or a distortion of its shape. Due to these factors, the brightness of the star varies as seen from Earth since different amounts of the surface at its normal brightness are seen at different times. These variations are regular since they are related to a feature on the surface of the star that is relatively constant.

Ellipsoidal Variable

An ellipsoidal variable star is a star in a binary system whose shape is deformed due to gravitational forces between the stars. Since the two stars are close together, the gravitational pull of each star on the other is stronger at the near side. The distortion in shape (to an ellipsoid from a sphere) causes different surface areas to be visible from Earth as the star rotates. This causes the amount of light seen from Earth to vary. The smaller the surface area visible from Earth, the dimmer the star will appear.

BY Draconis

The BY Draconis type of variable star is a star with starspots (similar to sunspots, but on other stars) that causes its brightness to vary slightly (from a few 0.01 magnitudes to 0.5 magnitudes). These stars are usually spectral type G, K, or M, so they can be dwarf stars, or stars like our Sun. The periods range from a few hours to over 100 days. This type of variable can also be variable in brightness due to flares on the surface of the star.

Although these stars vary in brightness for intrinsic reasons, they fall under the category of rotating stars, which, in its most common form, is an extrinsic variable. However, these stars are intrinsically variable stars whose variation is due to the star's rotation.

FK Comae Berenices

This class of star is a fast-rotating giant star of spectral type G or K. These types of variable stars have variations that have the same period as their period of rotation. The variation in brightness of these stars is very slight (only a few tenths of a magnitude). These stars appear to be high-mass stars that have advanced to the giant stage in their evolution.

SX Arietis

This type of star is a main sequence B star with emission lines (helium and silicon) that varies in brightness. These stars also have strong magnetic fields.

The variation in brightness of the star is typically about one day, which is about the same as the rotation period of the star. The variation in brightness observed is usually about 0.1 magnitudes.

BINARY STARS AND MULTIPLE-STAR SYSTEMS

Binary stars and multiple-star systems are common in the universe. The process of star formation forms at least as many multiple-star systems as it does single stars, like our Sun. Like binary star systems, in multiple-star systems the stars orbit a common center of mass (called the **barycenter**). This can be observed in nearby systems of stars where both radial and transverse motions can be measured. To determine whether a system of stars is a binary or multiple-star system, it is necessary to observe the stars for a long enough time to verify that they are, indeed, gravitationally bound to one another.

The Lonely Sun

Our Sun is somewhat unusual, then. Not only because we live on a planet that orbits it, but also because it is a single star, with no other star gravitationally bound to it. Our Sun is *not* part of a binary or multiple-star system, like most of the stars in our Galaxy, and, presumably, in the universe. It is unusual to find a star that is not gravitationally bound to another star, but this is not a *rare* type of star, just not the most common type of star.

Of course, it could be that our Sun *was* a member of a multiple-star system and got kicked out. (That happens sometimes, but because of gravitational interactions, not because the other stars don't like it!) But, the characteristics of the velocity (speed and direction) with which our Sun travels around the Galaxy do not support this as a possible scenario. All the evidence points to a lonely existence for the Sun with no gravitationally bound partner for the Sun at any time in its existence.

•••

The Sun as a Unique Star

For a long time, astronomers thought that the Sun's lone star status was the reason it had planets surrounding it. This idea supports the philosophy that life in the universe is rare because if planets are rare, then the likelihood of a planet like Earth is even more rare, and the likelihood of life forming on a planet like a rare Earth could be unique. This school of thought is also called "anthropocentric" because it leads to the conclusion that humans are the only intelligent life form in the universe.

In general, this philosophy is distasteful to a field of study that has been foiled twice when the assumption was that the Earth or the Sun were unique objects located in a unique part of the universe or galaxy. The first astronomers thought that Earth was located at the center of the universe or solar system. (The early astronomical concept of universe included stars, but they were much farther away than the planets, Sun, and Moon, so models of the universe showed only the locations of the planets, Sun, and Moon relative to Earth.) That turned out to be incorrect. The Sun is at a focal point of each of the elliptical orbits of the each planet in the solar system.

After that, astronomers thought the Sun was located at the center of the universe or galaxy. (These astronomers did not know yet that the universe consisted of many billions of galaxies; they thought the Milky Way was the universe.) Heliocentrism (the belief that the Sun was the center of the universe) turned out to be incorrect. The Sun is just one of hundreds of billions of stars in our galaxy, the Milky Way. It is located in the disk of our spiral galaxy about halfway out from the center of the galaxy.

So, being burned two times, astronomers are loathe to hold the position that Earth and the Sun are somehow so special that there is nothing like them in the universe. Most astronomers would agree that it is possible that this is true, but not likely, given what we know about the universe.

Recent studies of exoplanets indicate that planets can form in multiple-star systems. Several of the known planets that orbit stars other than our Sun are orbiting stars that have gravitationally bound companions. The state of at least a few of these systems indicates further that planets can even survive post-main sequence evolution of nearby companion stars (which is discussed in great detail in this chapter). It appears that the formation of planets is not an unusual process at all.

The fact that no astronomer has yet detected evidence of an Earth-like planet in a multiple-planet system (like our solar system) does seem to support the anthropocentric point of view; however, astronomers who study exoplanets insist that the lack of evidence is due to their inferior ability to detect such small planets. It is true that planet finders are still working to enhance their methods for planet detection to be able to find smaller planets. Most exoplanets found to date are more like Jupiter than like Earth.

Binary Systems

Binary star systems are very common. As we discussed in the previous section of this chapter, there are many types of variable stars that are binary systems. This is the simplest gravitational interaction one can describe between stars: two stars rotating around a common point that is the center of mass of the system.

Simple as it may seem, there are still many kinds of binary star systems that astronomers have studied over the years. In large part, these "types" have to do with how they are observed, and not some physical property of the systems that makes them unique.

Visual Binaries

The first type of binary star system is the visual binary star system. This is a pair of stars that appear to be orbiting one another. This type of binary star system may or may not be physical. That is, the stars may be gravitationally bound, or not. If the stars are not gravitationally bound and do not orbit one another, the pair is referred to as a "double star" to indicate the lack of physical connection between the stars.

An example of a double-star system is Alcor and Mizar in the constellation Ursa Major. This double-star system is actually visible to the naked eye to some. The second star in the handle of the asterism known as the Big Dipper is actually two stars. Some individuals can separately detect the two stars without the aid of binoculars or a telescope. Ironically, Mizar, the brighter of the two stars when viewed with the naked eye, is actually a physical binary star system. That is to say, there are really *three* stars in the place of the second star in the handle of the "Big Dipper"!

Mizar A and B are one example of a physical binary system. Another example of a physical binary system that is also visual (but requires a telescope to observe) is the Castor binary system. The star Castor is one of the two bright stars in the constellation Gemini that make up the asterism known as the Winter Ellipse. As you look south in the winter night sky, Gemini appears on the upper left-hand side of the Winter Ellipse, and Castor is the bright star in Gemini that is higher above the horizon and part of the Winter Ellipse. This star is actually a physical visual binary system. The two stars appear as far as 5 arcseconds apart.

The Castor physical binary system is so far apart that a full orbit of the primary star has not been observed yet. Observations of this system date as far back as the late 18th century. In over 200 years, the companion has not completed a full orbit around the primary star in the system.

Astrometric Binaries

Physical binaries may include a star that is too faint to be detected. Alternatively, the binary system may include a compact object, like a black hole, that does not radiate any electromagnetic energy. In such cases, it is still possible to tell that there are two masses orbiting a common center of mass. If only one star is visible, careful observations of the star's position can reveal an invisible partner.

Because both stars are orbiting a common center of mass, both stars are moving. No matter at what angle this system is observed from Earth, there will be transverse motion of the visible star, as long as the system is close enough to Earth to measure such motion.

The process of careful measurement of positions of stars is called astrometry. Since these systems of stars can only be identified by such measurements, they are called astrometric binary star systems. Since both members are orbiting a common point in space (their center of mass), this type of star system is different from visual binaries, as all astrometric binaries are physical, whereas only some visual binaries are physical. Additionally, in visual binary systems, two stars are observed; whereas, in astrometric binary systems only one star is observed.

An example of an astrometric binary system is Sirius A and B. Only using special instruments called stellar coronographs can the companion

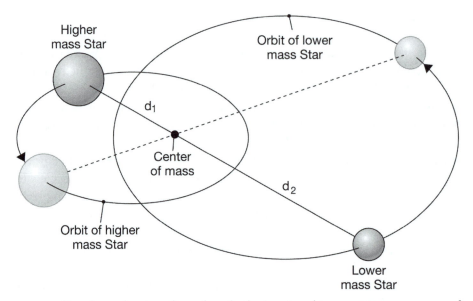

Figure 4.2 The above drawing shows how both stars in a binary system move around a common center, rather than the common idea of one star moving around another. [Jeff Dixon]

star, Sirius B, be observed. Sirius is the brightest star in the southern hemisphere. It is one of the bright stars that make up the asterism called the Winter Ellipse. Looking south, Sirius is the star at the lowest point in the Winter Ellipse. It is the brightest star in the constellation Canis Major (also known as the Big Dog).

The presence of the companion star, Sirius B, was inferred from the careful measurement of the position of Sirius A over many decades. The primary star seems to "wobble" with regularity. This is the signature of an astrometric binary star system. Not long ago, with the Hubble Space Telescope some of the first images of Sirius B were taken.

Spectroscopic Binaries

In addition to astrometric binaries, there are other scenarios where a binary system may look as though it is only a single star. For binary systems that are too far away to detect any transverse motion (at distances greater than a few hundred parsecs distant, changes in positions of stars result in angular differences too small to measure), there is another possible motion that can be measured. This is the radial motion.

Just as in astrometric binaries, spectroscopic binaries are systems of stars that are orbiting a common center of mass. As a result of this motion, the stars appear to move both back and forth (radially) and from side to side (transversely). In this case, alignment is an issue. There are some orientations a binary system could have that would not include any radial motion as seen from Earth.

However, for those binary systems that are oriented so that at least a portion of their motion around their center of mass is radial motion as seen from Earth, astronomers can measure their radial motion and identify that the star system is a member of a binary system. The radial motion is detected because of the Doppler Effect, which causes a **Doppler shift** in the wavelength of light observed. Just as the pitch of the sound of a train changes as it passes you from high pitch to low pitch, the light emitted from an object moving *away* from Earth is shifted to *longer* wavelengths. Conversely, the light emitted from an object moving *towards* Earth is shifted to *shorter* wavelengths. So, to distinguish this system as a binary star system, astronomers have to observe its spectrum.

For this reason, these systems are called spectroscopic binary star systems. A spectroscopic binary is one that cannot be observed visually or measured astrometrically. A spectroscopic binary is a system consisting of two stars that are gravitationally bound to one another and orbit a common center of mass. The only way to determine that the single star that is observed is actually two stars that are physically connected to one another is to observe and analyze its spectrum. There are two different kinds of spectroscopic binaries: single line and double line.

Single-Line Spectroscopic Binaries

Sometimes the spectrum of the star that is observed exhibits only a single set of lines corresponding to one spectral type. These lines, however, appear to shift back and forth over time. The shifting of the lines indicates radial motion. This is similar to the astrometric binary system in that the observation is of a single star (or stellar spectrum) that exhibits motion indicating the presence of another object. The other object may not be visible or may not have a distinct spectrum, but its presence is evident in the motion of the star or stellar spectrum.

Double-Line Spectroscopic Binaries

The more obvious type of spectroscopic binary is a system that is visually observed as a single star, but the spectrum indicates that there are two stars. The spectrum of such a system appears to be the spectrum of two different types of stars in one. The lines for both stars may shift or not. Usually, it is assumed that even though the lines do not shift, the system is still a physical binary system; however, it is possible that the observed double-line spectrum merely indicates a chance alignment of stars. In this case, the system is more like the visual binary system. It is not always true that the double-line spectroscopic binary system is a physically gravitationally bound system of stars.

Eclipsing Binaries

This configuration of binary stars was briefly discussed in the previous section of this chapter. Eclipsing binary stars are stars that are in a gravitationally bound system and orbit a common center of mass and whose alignment with respect to Earth is nearly edge-on so that one star passes in front of the other as seen on Earth. When the dimmer (cooler) star is in front, the brightness of the pair decreases most (primary minimum). When the brighter (hotter) star is in front, the brightness of the pair decreases the least (secondary minimum). When both stars are visible from Earth, the brightness of the pair is at its maximum.

Eclipsing binaries are very important in astronomy because they are the only systems that allow for the measurement of a star's mass and radius. A visual astrometric binary system will allow for a measurement of each star's mass. Spectroscopic binary systems allow for the direct measurement of the speed of the stars when they are moving radially with respect to Earth. But, without knowledge of the separation between the stars, the inclination of the system with respect to Earth, or the distance to the stars in question, only the mass of the *system* (not the individual stars) can be inferred from that information.

With eclipsing binary systems, the inclination of the system is limited to a very narrow range, so estimates can be made without knowing the exact inclination. Hence, eclipsing binary systems that are not also visual astrometric or spectroscopic binaries are still useful. However, to determine a mass of an eclipsing binary system, the distance to the system must be known. The speed of the stars and the diameter of each star are measured from the light curve.

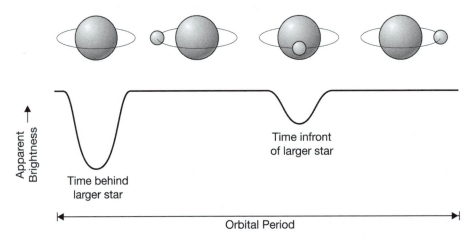

Figure 4.3 The above diagram shows how the brightness of an eclipsing binary star system changes as the two stars pass in front of one another. When the brighter star is in front of the fainter star, the peak is not as low as when the fainter star eclipses the brighter star. When both stars are visible, the brightness is at a maximum for the system. [Jeff Dixon]

The primary minimum is the deeper dip in the curve; this is when the cooler (dimmer) star is in front of the hotter (brighter) star. A portion of the light coming from the brighter star does not reach Earth, instead, the light coming from the dimmer star does. The secondary minimum is the shallower dip in the curve; this is when the hotter (brighter) star is in front of the cooler (dimmer) star. In this case, the light coming from the dimmer star does not reach Earth, instead the light coming from the brighter star does. The brightness maximum is the sum of the brightnesses of the two stars since light from both the dimmer and the brighter star is reaching Earth.

Since the x-axis of this curve is time, it is possible to measure the time it takes for one orbit to occur. Knowing the distance to the system and assuming circular orbits, it is possible to determine the ratio of the masses of the two stars. (If the angular separation of the stars is measurable, the mass of each star can be determined.) Using the light curve, knowing the time it takes for the dimmer star to pass in front of the brighter star, and vice versa allow for the calculation of the linear diameter of each star. (If the light curve shows flat-bottomed minima, then the actual diameter of each star is measurable, otherwise the distance measured is only some fraction of the diameter of each star. In other words, a minimum or lower limit to the diameter is measured.)

Multiple-Star Systems

Like binary star systems, multiple-star systems (systems containing three or more stars) can be physical or not. Systems that are gravitationally bound are called *physical* multiple-star systems. Systems that are not gravitationally bound, but are chance alignments of nearby stars with more distant ones are called *optical* multiple-star systems. Once a physical multiple-star system includes about 100 stars, it is dubbed an open cluster (open clusters are discussed in detail in chapter 3).

Open clusters have significantly different stellar dynamics than does a multiple-star system. In the case of multiple-star systems, the orbits are hierarchical, so that smaller orbits are nested completely within larger orbits so that they do not interact and remain stable. A three-body system where the orbits interact with one another always results in one body being expelled from the system. Only a two-body system is stable indefinitely. Multiple-star systems typically contain a two-body system with a third star in orbit around the binary system. More complex stable systems have been observed, but they consistently employ hierarchical orbits.

Open clusters, on the other hand, have complex stellar orbits that interact in chaotic stability. Nonlinear dynamicists study these kinds of systems where large numbers of gravitationally attractive masses interact with one another, yet stay together as a cohesive gravitationally bound system. These are complex mathematical endeavors that are beyond the scope of this text and most undergraduate college courses.

Coevolving Star Systems

In some star systems, at least two stars are close enough to one another that when one expands to become a red giant or supergiant, some mass can be transferred to the other. To understand how this can happen, we must first understand something about the gravitational field between these two orbiting bodies. It is, perhaps, easy to think about the gravitational attraction between two bodies along a line that connects them.

Assuming we are talking about two stars with the same mass, as we go along this line from one star to the other, the gravitational pull of the nearer star dominates. In this case, at a point exactly between the two stars, the pull of each is the same. At this point, a test mass would feel a net force of zero because both stars would be pulling equally on that mass in exactly opposite directions. This point is called a stable point. An object could remain at this point in space (exactly half way between two equal masses) indefinitely. If the object got just a smidge closer to either star, it would be pulled toward that star, however. This kind of stable point, where a small perturbation results in an object losing stability, is called a "quasi-stable point." It is sort of like a spherical rock balanced on the peak of a hill. Any push in any direction will send the rock down the hill.

Now, consider a system that is more realistic: the masses of the two stars are *not* equal. In this case, the stable point will not be exactly half way between the two stars, but somewhere else. Where? Well, if we consider that the more massive object exerts more force on a test mass over a greater distance, then the point of stability must be closer to the lower mass object. This can actually be calculated using the Universal Law of Gravitation:

$$F = Gm_1m_2/r^2$$

where G is the Universal Gravitational constant (6.67×10^{-11} Nm^2/kg^2), m_1 is the mass of the star (in kg), m_2 is the mass of the test mass (in kg), and r is the distance between the mass and the star (in meters). To solve for the position where the net force from both stars is zero, one would call the distance from star 1 to the equilibrium point, x, and the distance from star 2 to the equilibrium point, $r-x$, (where r is the distance between the two stars), then set the force due to star 1 on the test mass equal to the force due to star 2 on the test mass and solve for x.

This calculation will reveal that the stable point between two stars of different masses will be closer to the star with the lower mass. This seems a little counterintuitive, perhaps, but if you consider that the more massive star is pulling harder on the test mass, then the test mass has to get farther from the more massive star to find the equilibrium point where the force from each star is balanced by the force from the other star.

It turns out that there are other points of stability around a two-body system. There are five significantly stable points, called Lagrangian points, after the mathematician Joseph-Louis Lagrange. These points are locations around the two bodies where a test mass could remain indefinitely.

When we consider two stars that evolve, changing size, it is easy to see how matter transfer can occur. All that has to happen is for one star to expand beyond the stable point. Then, material from the expanding star will be pulled toward its companion. If material from one star gets pulled towards its companion, it will dramatically change how both stars will evolve.

Since massive stars evolve more quickly than low-mass stars, this scenario would only occur if the two stars are close enough together that the more massive star will expand beyond the stable point during its expansion to the red giant phase. In such a case, the material that will fall on the lower mass star will be mostly hydrogen, since massive stars have large radiative zones in their outer layers and much of that material will never be converted into helium through fusion.

So, the lower mass star will accumulate an amount of hydrogen on its outer layers. That will change the stability of the lower mass star dramatically. The extra mass will cause the star to collapse a bit more, since the fusion rates inside were not necessarily meant to support such a great mass. This mass transfer would cause the lower mass star to begin to evolve as if it were a higher mass star. Depending on the initial mass of the lower mass star and the amount of material it accreted from the higher mass star, it could potentially cause a star that was destined to become a white dwarf to go supernova and leave behind a neutron star or black hole.

For the more massive star, this change in mass would not likely change the outcome of its evolutionary cycle, but it could slow it down or speed it up, depending on how much mass the star originally had and how much was lost to the companion. Essentially, the difference would be that the core would not have to generate as much radiation to support the outer layers, since they wouldn't be there anymore. That might reduce the collapse of the core and the fusion rates. Conversely, losing the outer layers could cause the core to be more exposed causing a greater amount of hydrogen fuel to be available to the outer (shell) layers of the core resulting in an accelerated evolutionary sequence.

Close binary systems where mass is exchanged are not very common. They are an interesting laboratory for studying the intricacies of stellar evolution in depth. Unfortunately, the exchange of mass between stars does not cause stellar evolution to occur fast enough to observe on human time scales, so much of what we know is based on models with some observations to support the conjectures.

RECOMMENDED READINGS

Asimov, Isaac. *Asimov on Astronomy.* New York: Bonanza Books, 1988.
Bennett, Jeffrey D., Megan Donahue, Nicholas Schneider, and Mark Voit. *The Cosmic Perspective.* 5th ed. San Francisco: Benjamin Cummings, 2007.

DeGrasse Tyson, Neil, Charles Tsun-Chu Liu, and Robert Irion. *One Universe: At Home in the Cosmos.* Washington, DC: Joseph Henry Press, 1999.

Freedman, Roger, and William J. Kaufmann III. *Universe.* 8th ed. New York: W.H. Freeman Company, 2008.

Hawking, Stephen. *On the Shoulders of Giants.* Philadelphia: Running Press, 2002.

Jastrow, Robert. *Red Giants and White Dwarfs.* New York: W.W. Norton and Company, 1990.

Seeds, Michael A. *Astronomy: The Solar System and Beyond.* 5th ed. Pacific Grove, CA: Brooks Cole, 2006.

5

What Makes a Galaxy?

At some point in your life, if you haven't done it already, you should go to a dark location (far from city lights) on a clear summer night and look at the Milky Way. Immediately, you will understand why it is named the Milky Way, since it looks as if someone poured a large container of sparkly milk across the night sky. That's our Galaxy, the Milky Way. In fact, the word "galaxy" comes from the Greek word for the Milky Way (*galaxias*), which is derived from the words for mother's milk (*gala*) and circle (*kuklos*). The Greek meaning has to do with a myth about how the Milky Way was formed.

In 1609, Galileo was the first person to record an observation of the Milky Way showing that it was made up of many stars. Before this observation, it was believed that the Milky Way was a single cloud-like structure that spanned the heavens. There was speculation that it might be made of gases. No one, however, thought that it might actually be billions of stars.

So, is a galaxy billions of stars? Well, it's a little more than just stars. A galaxy contains stars, gas, and dust. All these materials must be gravitationally bound together to make up a galaxy. The remainder of this chapter will discuss the constituents of a galaxy: gas, dust, and stars, and their many forms within a galaxy.

GAS IN GALAXIES

The gas found in galaxies is nothing like the gas you put in your car; it is more like the air you breathe. We call the gas found in galaxies interstellar gas. Interstellar just means between the stars. Gas (along with dust) in galaxies is found in between the stars. In the case of interstellar gas, the word "gas" describes a state of matter—the gaseous state. Most interstellar gas is hydrogen in its atomic form. That means the gas is a cloud of atoms of hydrogen. Some interstellar gas is in molecular form (a molecule is at least two atoms bound together by electronic bonds). It is in these **molecular clouds** that dust is often found.

States of Matter

Matter comes in four known states: solid, liquid, gas, and plasma. Most of us have significant experience with the solid and liquid states of most matter. Some of us have significant experience with gaseous matter (beyond just breathing it!). Almost none of us have significant experience with the plasma state of matter, but it is most important in astronomy as much of the hot gas in outer space is in the plasma state.

Neutral Atomic Interstellar Gas

Primarily, neutral atomic interstellar gas is hydrogen. Since the gas is neutral, this means that each hydrogen atom consists of one proton and one electron. So, each atom has a net charge of zero (since the proton has a positive charge of one and the electron has a negative charge of one).

Typically, in astronomy, the Bohr model of the atom is used as a reference. The Bohr model describes an atom as a nucleus with electrons in orbit around the nucleus. The orbits in the Bohr model are circular. This is where the model is inaccurate—actually, the electrons move in quasi-chaotic orbits around the nucleus and the orbits have several different shapes that define them. For our purposes, though, the Bohr model is complete enough. The shape of the orbits is irrelevant to our discussion. Rather, what is important is that each orbit represents a different energy state for the electron. In this aspect, the Bohr model is consistent with modern atomic theory. Figure 1.12 in chapter 1 of this volume shows a diagram of the Bohr model of a hydrogen atom.

Each energy level in the atom, as defined in the Bohr model, defines an "allowed" state for the electrons in that atom. In the case of hydrogen, there is only one electron, but the different orbits still define energy states in which that electron can exist. In a hydrogen atom, the electron cannot have an energy that is in between any two of the energy states.

When hydrogen is in the neutral form, the electron is usually at its lowest energy state (or **ground state**). That means the electron is in the smallest orbit around the proton. Neutral atomic hydrogen has two ground states. This is because both the electron and the proton are spinning. Since both components of the atom are spinning, as well as orbiting one another, there are two possible ground states. The higher energy ground state occurs when the spin of the electron and the spin of the proton are in the same sense (both particles spinning clockwise or counterclockwise). The lower energy ground state occurs when the spin of the proton and the spin of the electron are in opposite directions (one spinning clockwise and the other spinning counterclockwise). The transition from the higher energy ground state to the lower energy ground state is called a **spin-flip transition,** since the electron changes orientation so that its spin is opposite that of the proton.

The spin-flip transition is a change in energy equivalent to the energy of a photon with wavelength 21cm. Neutral atomic hydrogen gas, therefore, emits electromagnetic radiation at a wavelength of 21cm. Through observations of this emission line, astronomers can study neutral atomic hydrogen. Observations of neutral atomic hydrogen in the Milky Way tell us about the rotation of the Galaxy. Observations of 21-cm radiation in other galaxies tell us about the mass contained within those galaxies as well as the rotational motion of those galaxies (galaxy rotation and how it is related to mass are discussed in chapter 7 of this volume).

Ionized Atomic Interstellar Gas

Ionized atomic interstellar gas is mostly hydrogen, but a significant portion of ionized atomic gas in galaxies is *not* hydrogen. Other elements that astronomers study in their ionized states include: oxygen, nitrogen, iron, helium, neon, carbon, sulfur, and many more. A gas is ionized when one or more electrons are removed from the neutral atom.

Typically, gas is ionized through radiative processes. That is, electromagnetic energy is the mechanism that frees electrons from neutral atoms making them ionized. This often occurs around the hottest stars because the majority of the electromagnetic energy produced is in the form of high energy photons that easily ionize many atoms. Recall that, in the Bohr model of the atom, electrons exist in specific energy states. Each energy state defines the amount of energy needed to free the electron from the atom. Also, recall that in the Bohr atom, each energy state marks an "allowed" state for the electron. For an electron to change energies, it must move only from one energy state to another. There is no "in between" space for electrons. (Think of walking up or down a staircase. There is not really a way to be "in between" stairs; at least one of your feet is always on a stair surface.) Each element's energy diagram is different and specific to that element, so that the energy diagram for hydrogen is not like the energy diagram for any other element.

In addition to radiative ionization processes, there are also mechanical ionization processes that occur in the interstellar medium. Most often we see this type of ionization in areas where gas is moving very fast. For example, near a recent supernova event, there might be an expanding shell of ionized gas. Or, near a **supermassive black hole,** there might be gas ionized due to the high-speed collisions of gas particles in clouds and shear between clouds of gas within an accretion disk being consumed by the black hole. These incidents of mechanical (sometimes called collisional) processes that ionize interstellar gas are usually indicative of unusual physical conditions.

Ionized gas is most often studied by observing the emission spectrum of the gas. This has historically been done primarily at visible wavelengths. However, in the last few decades new instruments have been developed to detect radiation in the ultraviolet, infrared, and microwave regimes; these instruments make it possible for astronomers to study more of the emission spectra of astronomical objects than ever before. As a result, astronomers are learning more about the kinds of ionization processes that go on in the interstellar medium.

Plasmas can be created in laboratory situations and studied using at the same wavelengths that astronomers study the interstellar ionized gas. Based on the laboratory studies, astronomers can learn about the density and temperature of interstellar ionized gas. In the laboratory studies, different transitions of the electrons within an ionized atom will be more or less frequent depending on the ionization or excitation method and the density and temperature of the ionized gas. So, astronomers can take the laboratory data, as well as the theoretical predictions for scenarios that cannot be recreated in a laboratory, and compare these to what they observe in the interstellar medium.

Molecular Interstellar Gas

Interstellar gas in molecular form is mostly molecular hydrogen gas (H_2). The second-most abundant and well-studied interstellar molecular gas is carbon monoxide (CO). There are plenty of other, more complex, molecular gases found in the interstellar medium, but they are far less abundant and remain difficult to understand.

Molecular hydrogen, (H_2) as it turns out, cannot be studied directly since it does not radiate significantly at any observable wavelength. (Except for a collisionally excited emission line found in the near-infrared regime, there is no well-studied emission from H_2.) This has to do with the structure of H_2. Two hydrogen atoms joined together have no dipole moment. That is, the molecule has no preferred orientation in the presence of an electric, magnetic, or gravitational field. Therefore, the presence of any of these fields will not cause the atom to lose an electron, rotate, or vibrate, so the electrons will not change energy states and no electromagnetic radiation will be emitted. Emission from H_2 will only occur if the hydrogen molecule is caused to vibrate or rotate by some mechanical process. This appears to happen only in extreme situations, like near supermassive black holes or recent supernova events.

To study H_2, astronomers have studied extensively CO emission in the radio regime. CO seems to always be associated with molecular hydrogen. Astronomers have even tried to derive a relationship between the amount

CO detected and the amount of H_2 associated with the CO. The relationship is still not one that is widely agreed upon. Different astronomers use different methods to derive the relationship, and many different answers are considered acceptable. (Much like the value of the Hubble Constant in cosmology. For more information on the Hubble Constant and its use in cosmology, see the *Cosmology and the Evolution of the Universe* volume of this series.)

The primary method for the production of electromagnetic radiation from molecular gas is through rotation or vibration of the molecule. Since both of these processes involve accelerating a charge or group of charges, electromagnetic radiation is produced. These processes involve very low energies, so the radiation produced is often at very long wavelengths. Interstellar molecular gas, therefore, is primarily studied in the radio regime.

••

Molecules

Most molecules are fairly fragile. The loss of an electron usually means the destruction of the molecule. Once a molecule loses an electron it is either an ionized molecule (unbalanced charges), another molecule (because it also loses a nucleus), or just individual atoms. Sometimes, if the molecular bond is broken and leaves behind a different molecule, the new molecule can be studied (like H_2C_6 breaking down into H_2 and three molecules of C_2). More often, though, the breaking of a molecular bond is due to some strong radiative process that will likely break any newly formed molecules and ionize any atoms.

There are many different molecules found in space. Most are carbon based, due to the fact that carbon creates the most stable molecular bonds. It is thought that all life in the universe ought to be carbon based because of this fact and the fact that there is so much carbon in the universe.

Interestingly, the molecules found in space are fairly complex. One might think that the radiation from active stars and active galaxies and supernova remnants would make it difficult for complex molecules to form, but astronomers have found sugar (poly-carbon structures, like glycolaldehyde, $H_2COHCHO$) in our own Galaxy. In fact, since sugars are strings of carbons with hydrogen atoms attached, one could say (and be completely correct) that the Milky Way contains carbohydrates!

••

DUST IN GALAXIES

The second of the three main constituents of a galaxy is dust. Interstellar dust ranges in size from large molecules to small dust particles (a few hundred microns in size). Long molecular chains of hydrocarbons, called **polycyclic aromatic hydrocarbons (or PAHs),** are the smallest particles of dust. Large dust grains are typically a few hundred microns (micrometers) across (less than 1 mm) and may have layers of water ice on them. It is thought that most dust grains are either made of carbon or some kind of silicate.

It is difficult to study the composition of dust directly since the only observable features are the blackbody spectrum of the solid particles and

··

Kinds of Ice

In astronomy, it is often necessary to differentiate among different kinds of frozen gases and liquids. So, when astronomers talk about normal ice, they will usually say "water ice." This is because there are many other kinds of ice in the universe that astronomers can study. Even in our solar system, there are ices of methane, ethane, and ammonia in the atmospheres of the giant planets and in the compositions of some comets.

··

the reflected and scattered light due to the presence of the solid particles. The blackbody spectrum gives astronomers a way to determine the temperature of the dust. From these data, astronomers can hypothesize how the grains get heated to such a temperature and then build models to help narrow the possible chemical composition of the grains. The reflected and scattered light due to dust give astronomers some clues about the location of most dust as well as possible sizes and even chemical compositions of the dust grains. Obviously, there must be a lot of hydrogen in the dust grains (since hydrogen is the most abundant element in the universe), but some other element is required, chemically speaking, to create solid grains.

Polycyclic Aromatic Hydrocarbons

Interstellar PAHs are a fairly recent discovery in the pursuit of understanding interstellar dust. In the advent of infrared astronomy, which occurred in the early 1980s, observations of PAH emissions were numerous. On Earth, PAHs are considered air pollution. The Centers for Disease Control describes PAHs as "chemicals that are formed during the incomplete burning of coal, oil, gas, garbage or other organic substances" (Centers for Disease Control, n.d.). Since PAHs are so prevalent on Earth, they have been well studied, and their emissions due to radiative and collisional processes are well documented. There are many emission lines due to PAHs observed looking out into space from Earth in many directions (these detections are *not* due to **terrestrial** atmospheric PAHs). These observations support the assertion that interstellar PAHs must exist.

Theories have been developed to explain how PAHs might form in the inhospitable environment of outer space. Interestingly, PAH features are not *everywhere,* but rather appear to be associated with specific features in our Galaxy as well as other galaxies. PAHs appear to be most prominent in and around parts of the interstellar medium where stars are forming (so-called **star-forming regions**). It is this fact that casts a shadow of doubt on the possible existence of PAHs in the interstellar medium. Near such star-forming regions, the electromagnetic radiation from newly formed stars should be great enough to ionize and dissociate (tear apart) these molecules. Yet, they appear to exist. What's more, observations of these molecules are almost *exclusively* found in regions where they should not be able to exist.

The emission seen from PAHs, in the near-infrared and mid-infrared parts of the electromagnetic spectrum, is due to radiative processes. It was only after extensive studies and laboratory experiments designed to recreate environments similar to that of the interstellar medium that astronomers were able to demonstrate that, in fact, the previously unidentified emission lines in the infrared were due to interstellar PAHs.

Now, astronomers understand interstellar sources of near- (1–5 μm), mid- (5–50 μm), and far- (50–200 μm) infrared radiation far better than before, of course. Decades to study the emissions observed in these regimes, led to more theoretical work that began to explain the observations. Near-infrared radiation is primarily ionized atomic gas emission due to radiative and collisional processes. Mid-infrared radiation is largely dominated by PAH emissions. Far-infrared radiation is primarily blackbody radiation from interstellar dust grains with a smattering of emission lines of rare, highly ionized molecular and atomic gases.

Carbon and Silicon Dust Grains

It was hypothesized that the Galactic interstellar medium contained primarily carbon dust grains, specifically graphite dust grains. This is because of a bump in the interstellar **extinction** curve. The interstellar extinction curve tells astronomers how different wavelengths of light are extinguished (absorbed, scattered, and reflected) as they pass through the, presumably, uniform **interstellar medium.** To create this extinction curve, astronomers observe the spectra of hundreds of thousands of stars. Then, astronomers observe these same stars at every wavelength possible and then plot the amount of extinguished light for each wavelength observed.

In the late part of the 20th century it was discovered that this curve was remarkably similar in almost every direction within our Galaxy. Astronomers interpreted this to mean that the dust in our Galaxy is basically uniform. It is difficult to determine what interstellar dust is made of since interstellar dust, by definition, does not exist on Earth or even within the solar system. Astronomers decided to analyze the extinction curve to try to determine what interstellar dust is.

The composite extinction curve shows a distinctive "bump" at about 217.5 nm. Laboratory experiments with different materials narrowed the composition of interstellar dust grains to a small number of possibilities. Following these laboratory experiments, theorists began to fit the observed extinction curve with the different materials, making some assumptions about grain sizes. The optimal fit seems to include at least two types of grains: graphite grains and silicate grains. The sizes of these grains appear, now, to vary considerably from region to region with larger grains in cold, dense clouds and smaller grains in the hot diffuse interstellar medium.

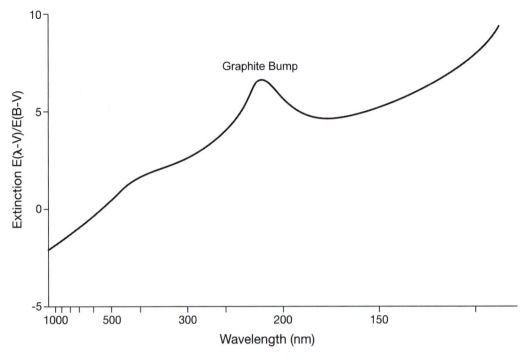

Figure 5.1 This diagram demonstrates the Bohr atomic model. In this model the nucleus of the hydro-gen atom consists of one proton. Orbiting around the nucleus is one electron. The electron is pro-posed to have several allowed orbits located at different distances from the nucleus. Each orbit is proposed to be circular. This model is not physically accurate—the electron does not orbit the proton in a circular orbit at a fixed distance from the proton. However, despite this inaccuracy, the model is useful for visualizing the changes in energy levels that the electron does undergo when photons are absorbed or emitted. [Jeff Dixon]

The evidence that supports these claims is complex and difficult to com-municate without first giving a laundry list of all the features of the extinction curve and newly observed emissions of the interstellar medium; however, it is important to note here that the extinction curve for our Galaxy is *not* the same as the extinction curve for other galaxies (however, it is surprisingly similar to the extinction curves of the most actively **starbursting galaxies**). In fact, it is well known that most other galaxies probably have different dust constituents and size distributions. For example, two of the galaxies nearest to ours (the Large and Small Magellanic Clouds) do not appear to have the graphite "bump" feature, and probably, therefore, do not contain as much dust that is carbon based.

STARS IN GALAXIES

The third and final constituent of a galaxy is the stars. Most galaxies con-tain a huge number of stars (several billion), although some have as few as

10 million. Even with these large numbers of stars, stars do not dominate the masses of galaxies. Of the three constituents described in this volume, gas (specifically, molecular gas) is the greatest contributor to the mass of a galaxy. The populations of stars that make up a galaxy reflect the galaxy's star formation history.

Galactic Mass

In fact, *none* of the constituents of a galaxy mentioned so far dominate the mass of a galaxy. The dominant contributor to the mass of a galaxy is something known as dark matter. The discovery of dark matter, as well as current hypotheses about its nature will be discussed in detail in chapter 6.

Some galaxies contain both old and young stars. These galaxies have been forming stars more or less continuously since the galaxy was formed. Many galaxies like our own Galaxy are like this, containing both older and younger populations (including the highest mass main sequence stars), as well as active star-forming regions where objects that are not yet considered stars exist.

Other galaxies contain only old stars (no main sequence stars with lifetimes less than 1 billion years, for example). Such galaxies stopped forming new stars some time ago (1 billion years ago, in the example above) so that the only stars in the galaxy are stars that formed before the time when stars ceased to be able to form. Of course, in these galaxies, there are no active regions of current star formation. In fact, we also find that in these galaxies the ingredients needed for star formation are not present in the necessary proportions.

Still other galaxies contain only a few stars and are predominantly made of gas and dust. These extremely faint galaxies are fairly new discoveries (in the late 20th century), although some astronomers hypothesized that such galaxies might exist decades before they were discovered. These **low surface brightness galaxies** also appear to have the ingredients for star formation unbalanced. Since these galaxies are such new discoveries, they have not yet been studied well enough to determine whether these galaxies recently formed stars or stopped forming stars a long time ago.

GALAXY SIZES

Galaxies, like stars, come in a range of sizes and shapes. In this section of this chapter, we will explore the range of sizes that known galaxies exhibit. There may be a wider range of galaxy sizes than what is discussed here, but in the early 21st century, this encompasses what is known about the possible range in size of galaxies.

The Smallest Galaxies

The smallest galaxies are barely larger than a cluster of stars. In fact, it may be possible that some of the objects currently called **dwarf spheroidal galaxies** may be runaway giant globular clusters. The smallest known galaxies were discovered just a few years ago in a nearby cluster of galaxies, known as the Fornax cluster. A team of Australian astronomers took images and spectra of these newly named **ultra compact dwarf galaxies.**

Ultra compact dwarf galaxies contain a few tens of millions of stars and are only a few tens of parsecs across. Yet, these tiny compact systems of stars contain all the constituents of a galaxy. Their extremely small size makes them about 10,000 times less luminous than the average galaxy.

Astronomers are still trying to understand whether these galaxies are the building blocks of larger galaxies or whether they are the tiny cores of small galaxies that have been stripped of their discs through some violent interaction. Current theories of galaxy evolution suggest that there should be more small galaxies than large ones, as our current understanding of the formation of the universe and the evolution of galaxies involves building large galaxies from small pieces. Thus, there should be lots more of the small pieces of galaxies still available to build larger galaxies. The recent discovery of ultra compact dwarf galaxies may turn out to be new evidence to support the current paradigm of galaxy evolution.

The Average Galaxy

The average galaxy contains close to a billion stars, along with the normal proportions of gas and dust in their many different forms. Usually the gas and dust are located only in a disc, while the stars are spread more uniformly. However, not all galaxies are disc dominated (as we will see in more detail below), so sometimes the gas and dust are more spherically distributed. The average galaxy has an absolute magnitude of -20. This means that the average galaxy has a luminosity of about 10^{40} erg/s or 10^{34} Watts.

Magnitude Scales

The magnitude scale is a logarithmic scale of brightness. There are two parallel such scales in use—absolute magnitude and apparent magnitude. Absolute magnitude is a measure of the brightness of an object at a fixed distance from earth of 10 parsecs. Apparent magnitude is a measure of the brightness of an object at its actual distance from earth. To measure absolute magnitude, one must know or calculate or determine in some way the luminosity (energy per second) emitted at the surface of the star. (For more information on these scales see chapter 2 of this volume.)

The average galaxy is a few tens of thousands of parsecs in diameter. Even though galaxies can have many different kinds of shapes, astronomers often describe the size of a galaxy by its diameter so that irregularly shaped galaxies have a defined diameter only when an imaginary circle is drawn around them.

••

Parsecs

A parsec is the distance to an object that has a parallax of one arcsecond. This distance is the equivalent of about 3.3 light years, or the distance light can travel in 3.3 years, or about 3.6×10^{16} m. (For more information on the parsec, see chapter 2 of this volume.)

••

The Largest Galaxies

The largest known galaxies are the **giant elliptical galaxies.** These galaxies can contain trillions of stars. Giant elliptical galaxies are found in the centers of dense galaxy clusters and are thought to have formed through many mergers of smaller galaxies. Since giant elliptical galaxies have very little gas and dust, their evolution must have been violent enough to trigger massive star formation episodes that quickly exhausted all the fuel for making new stars. Still, such galaxies show many different epochs of star formation (distinctly different-aged populations of stars). This fact implies that such periods of massive star formation must have recurred periodically. The model for giant elliptical galaxy formation being due to many *different* mergers, therefore, is supported by this evidence.

Giant elliptical galaxies are close to 10 times brighter than average galaxies and can be up to 2 million parsecs across. This is larger than most **groups of galaxies** and very close to the size of many **clusters of galaxies.** As you might have guessed these galaxies are not irregular in shape at all, rather, they are quite regular and do not include a disc component to their structure.

GALAXY MORPHOLOGY

In the beginning of the 20th century what is now known as a galaxy was called a "nebula" (pronounced "neb-you-lah"). The word nebula means "cloud." In those days, astronomers thought that all the fuzzy objects they were seeing were the same thing—nearby clouds of gas and dust with a few stars illuminating them. Some of the things astronomers named nebulae (the plural of nebula; pronounced "neb-you-lee") turned out to be just what they thought—clouds of dust and gas; however, many of these objects were galaxies. In fact, each object astronomers called a "spiral nebula" was, in fact, a galaxy. Today, these objects are known as **spiral galaxies.** Spiral galaxies are just one of the many kinds of disk galaxies known.

Before astronomers even understood what galaxies were, the morphological classification of these objects was already underway. The morphology of a galaxy is a study of the galaxy's shape. The root "morph-" means "shape" in Greek. Since astronomers in the late 19th and early 20th centuries had telescopes powerful enough to see many galaxies, they began to classify what they saw by their shapes. Without understanding more than that there were many of these objects and they appeared to have different shapes, astronomers went about trying to understand them by separating them into groups.

This process is very similar to the process we all use to learn. First, we see something we don't understand, then we try to understand it by finding similarities and differences between the thing we don't understand and things we do understand. Finally, we learn enough about the new thing to be able to identify it as a thing different from everything else we know, but similar to many different things (perhaps) that we do understand. This is the beginning of the scientific way of knowing. Scientific knowledge requires consensus, or agreement, among all observers. So, when an individual sees something new, she should be able to describe it in terms of things she, and others, know and understand. That way, others who find the same thing will be able to identify it.

Galaxies are most often described by their shapes, with some emphasis on any special relevant feature(s). The first classification scheme, developed by Edwin Hubble (after whom the famous Hubble Space Telescope is named), implied that there was an order within the classifications indicating an evolutionary sequence. Hubble hypothesized that galaxies started out as ellipticals and flattened over time becoming spirals in the end. It is still hypothesized today that the different shaped galaxies are, in fact, fundamentally different from one another; however, there is still little evidence to support this claim, since galaxy evolution is not entirely pieced together enough to confirm this hypothesis. For all we know, at this point in time, all galaxies started out the same way, but over time, due to their environments, or due to some other factors we don't yet understand, they changed and evolved differently, ending up with the various shapes and sizes we now observe.

All that aside, according to the most recent and best understanding of galaxies, they can be divided into three different groups. These are **disk galaxies, spheroidal galaxies,** and **other galaxies.** Originally, when galaxies were classified, there were more than three morphological groups. There were spirals, barred spirals, ellipticals, irregulars, peculiar galaxies, and more. In this section of this chapter, these different types will be discussed, but rather than presenting a zoo of galaxy shapes, the morphological types have been grouped together to emphasize some of the newest advances in understanding galaxy evolution that have occurred in the last few decades.

Almost a century after galaxies were first grouped together by shape, astronomers have learned much more about galaxies and understand far more than just the shapes of galaxies. The content of galaxies varies with morphology. Astronomers have inventoried the stars in several of the nearby galaxies. They learned that the stellar content of galaxies varies with morphology. The variation in stellar content tells astronomers something about galaxy evolution when stellar evolution is taken into consideration. Gas and dust masses vary with galaxy morphology. Additionally, star formation rates can be estimated for many galaxies.

The dynamics of a galaxy varies with morphology. Astronomers have measured the speeds of the gas and stars in many different galaxies of many different types. It has been shown that disk galaxies (spirals and barred spirals) have completely different kinematics (motions) than spheroidal galaxies (ellipticals). All these things together demonstrate in a fairly clear way that Hubble's hypothesis cannot be correct. Galaxies do not start out spheroidal in shape and gradually flatten over time. Still, many astronomers hypothesize that there may be some kind of evolutionary progression somehow discernable through Hubble's original morphological classification. Although, there is no clear evidence to support this hypothesis, yet.

So much more is known today than 100 years ago that the old morphological classifications seem almost tedious to present. The fact that galaxies can be grouped by shape and still be interesting to discuss is intriguing. It certainly implies that the shape of a galaxy must be significant. It is quite possible that the century-old morphological classification scheme can yet teach us something new.

DISK GALAXIES

Disk galaxies are those galaxies that have a component of their constituents rotating in a two-dimensional **disk.** This type of galaxy is, perhaps, more common than one might think. Originally, it was not known that the galaxies classified as spiral and barred spiral were, in fact, disk galaxies. Astronomers hypothesized that spiral and **barred spiral galaxies** must be flattened into a disk-like shape, but this was largely based on observations of many galaxies with spiral features that were viewed from different angles. The supposition was that all galaxies with spiral features were pancake-like in shape, but seen face-on, rather than edge-on. The counter supposition also applied: those galaxies seen edge-on were supposed to have spiral features, although they were not observable from the apparent angle.

In the last century, astronomers have been able to acquire data that support these hypotheses. So the idea of putting spiral galaxies into the same category as flat galaxies was entirely justified, in the end. It is quite interesting that very simple suppositions have turned out to be correct, when discussing the universe. It is probably for this reason that Occam's Razor is such an important principle for astronomers, in particular.

Occam's Razor

Occam's Razor is the principle that states "entities should not be multiplied unnecessarily" (*Merriam-Webster OnLine*). A 14th-century logician named William of Occam is attributed with devising this principle. William was a Franciscan friar and used this principle to justify many different things, including his belief that "God's existence cannot be deduced by reason alone" Isaac Newton adopted Occam's Razor for his own purposes and rewrote it as "we are to admit no more causes of natural things than such as are both true and sufficient to explain their appearances" (Physics and Relativity FAQ, 1997).

A disk galaxy, in the most basic sense, has four main observable features: a disk, a bulge, arms, and a halo. The disk is the pancake-shaped part of the galaxy. It is thin and extends from the bulge outward. Within the disk is the majority of the gas and dust contained in the galaxy. The bulge is a spheroidal feature in the center of the disk galaxy. While there is some gas and dust in most bulges, a bulge is predominantly stars. The arms are bright features found within the disk usually beginning at the edge of the bulge and extending to the outer edge of the disk in a long arc. Finally, the halo is the spheroidal component that encompasses all of the other parts of a galaxy. The halo also contains gas, dust, and stars, but is predominantly comprised of dark matter. (The nature of dark matter will be discussed in much greater detail in chapter 6.)

Our own Milky Way galaxy is a disk galaxy. You can see this for yourself by observing the Milky Way on a dark summer night. What we can see of our galaxy from our vantage point is the disk. The disk of the Milky Way appears as a thin band of stars and gas and dust across the sky. You can see more of the Milky Way from the southern hemisphere. From there, the bulge of the Milky Way is visible. The Milky Way's bulge is small, indicating that it is a truly disk-dominated object.

Normal Spirals

Normal spiral galaxies are disk galaxies with round bulges in their centers. These bulges can vary in size from a very small fraction of the disk to so much of the disk it is almost difficult to tell there is a disk at all. These galaxies are called spirals because the dominant features observed in the disks of these galaxies are the so-called spiral arms. These are bright features that extend from the bulge out into the disk in a long arc. Where the spiral arms end is usually considered the end of the visible part of the disk.

The spiral arms are the most prominent feature of a normal spiral galaxy. The nature of this feature will be discussed in much more detail in chapter 7 of this book. For now, we will say just that the arms consist of the same con-

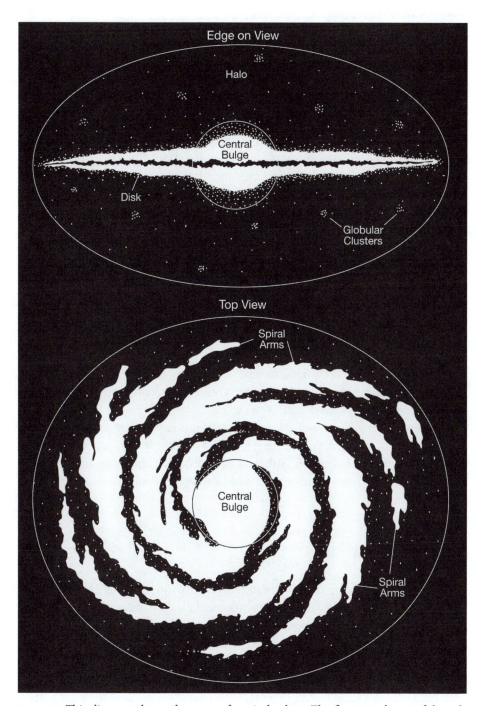

Figure 5.2 This diagram shows the parts of a spiral galaxy. The flat, round part of the galaxy is called the disk, the central, spheroidal part is called the bulge. The surrounding material that is both above and below the disk in a spheroidal distribution is called the halo. When observing the disks of some spiral galaxies face-on (so they appear round) one can see the spiral arms. These are long bright parts that extend from the bulge to the outer parts of the disk. [Jeff Dixon]

stituents of the galaxy: gas, dust, and stars. There are no more stars in the arms than in between the arms, but the types of stars in the arms are different from the ones between the arms. The stars in the arms are the youngest, most luminous stars in the galaxy. This explains, somewhat, why the arms are the most prominent feature of a spiral galaxy.

There are some spiral galaxies whose arms are not very prominent. Yet, the galaxies are still called "spiral." This is because of the historical nature of galaxy morphological classification. Originally, all disk galaxies were called spiral galaxies. Later, astronomers learned more about the nature of disk galaxies and many new morphological classifications became apparent. Now, we still call all disk galaxies spiral, but we have other adjectives we use to refine that classification. However, a spiral galaxy in the most general terms really means just that it's a disk galaxy, since not all spiral galaxies have prominent spiral arms.

Among the normal spiral galaxies, Hubble noted that the bulge size varied from large to small. So, he devised a scheme to classify galaxies by arm prominence, arm angle of separation, and bulge size. The symbol used for the classification of spirals is S. The subclasses are indicated by letters a–d. So, spiral galaxies can be Sa, Sb, Sc, or Sd, or anything in between (e.g., Sab, Sbc, etc.). The "a" galaxies have the largest bulges, while the "d" galaxies have the smallest bulges. Due to the attachment of Hubble's scheme to his hypothesis about galaxy evolution, Sa galaxies are often called "early type" spirals, and Sd galaxies are often called "late type" spirals.

As mentioned before, our Milky Way galaxy is a disk galaxy. Although it is not a normal spiral, the bulge size indicates that it is a "late type" spiral with a subclass of cd. The "cd" subclass indicates that the bulge of the Milky Way is not the smallest, but is a size in between the smallest and the next largest size.

Barred Spirals

Barred spirals are the first well-documented new morphology for spiral galaxies. Early in the era of galaxy morphological classification, it was observed that some spiral galaxies appeared to have elongated bulges. The center of the galaxy was not round, but bar-like, extending out into the disk. Even the arms seemed different, attached only to the ends of the central bar.

In the beginning, it was thought that there was some fundamental difference between so-called normal spirals with round bulges and the new barred spirals with elongated bulges. Today, it is understood that the bar feature is a transient feature in spiral galaxies. Astronomers are beginning to understand many of the different ways bars can form and sustain themselves in disk galaxies. Each of these methods must be long-lived, as the vast majority (66%) of all known disk galaxies are observed to have bars to some extent.

Figure 5.3 This image of NGC 1300 shows what a typical barred galaxy looks like. Notice that the arms of the spiral begin at each end of the bar, not on the sides or corners. [Credit: Hubble Heritage Team, ESA, NASA]

Our Milky Way galaxy is a barred spiral galaxy. The central bulge of the Milky Way is not round, but peanut shaped. The peanut shape is a common bar shape for small or weak bars. The unusual shape of the Milky Way's bulge was discovered in the 1990s. Without very high resolution infrared imaging (available only recently), it was difficult to demonstrate that the bar was indeed a bar and not an oval or some other elongated shape. More recent data from the Spitzer Space Telescope points to a very large and prominent bar at the center of the Milky Way.

Giant Spirals

While most spirals are average in size, there are several well-known giant spirals. The most well-known example of a giant spiral is the Milky Way. While average-sized spirals might have a billion stars, giants, like the Milky Way, have hundreds of billions of stars.

Usually giant galaxies have several satellite galaxies. Satellite galaxies are not exclusively around giant galaxies, but there are very few giant galaxies without known satellites. A satellite galaxy is usually a dwarf galaxy that is in orbit around another galaxy. Satellite galaxies are gravitationally bound to the giant galaxies they orbit.

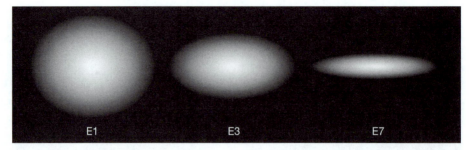

Figure 5.4 This diagram shows what an elliptical galaxy looks like. This shape is not the most flattened shape, but it shows a high aspect ratio so that one can see that one dimension is much larger than the other two. [Credit: NASA, ESA, and The Hubble Heritage Team (STScI/AURA)]

SPHEROIDAL GALAXIES

The other dominant shape for galaxies in the universe is spheroidal. These galaxies, notably, have no disks and have a shape that is three-dimensionally round. Originally, these galaxies were called **elliptical galaxies.** This name was given to these galaxies because scientists thought that they were seeing the same shape galaxy from different angles. So, when a galaxy appeared circular, it was assumed that this galaxy was elongated in the dimension corresponding to depth.

Spheroidal galaxies, as a group, contain much less gas and dust than disk galaxies. Some spheroidal galaxies appear to be almost completely devoid of these constituents. Spheroidal galaxies are the largest and smallest galaxies in the universe. Also, spheroidal galaxies are the most common type of galaxy in the universe. This is the case mostly because the vast majority of dwarf galaxies are spheroidals and there are far more dwarf galaxies than average-sized or giant galaxies in the universe.

Dwarf Spheroidal Galaxies and Ultra Compact Dwarfs

For a long time it was thought that dwarf spheroidal galaxies were the smallest galaxies in the universe. These galaxies contain a few hundred million stars and are usually three-dimensionally round, rather than cigar shaped, for example. Dwarf spheroidal galaxies are usually a few thousand parsecs in diameter. These galaxies are almost always found orbiting a larger galaxy. There are only a handful of dwarf spheroidal galaxies that have been found to be isolated.

In the beginning of the current century, a new, even smaller spheroidal galaxy was discovered. This new galaxy contains only 10 million stars and they are less than a hundred parsecs in diameter. The density of stars in these galaxies is so high that they have been dubbed "ultra compact galaxies."

These ultra compact galaxies were discovered in a well-studied system of violently interacting galaxies. Scientists think these objects formed in this hostile environment.

Elliptical Galaxies

Elliptical galaxies are so named because when it was discovered that nebulae (nebulae are what galaxies were called before they were known as galaxies) could be round, but not have spiral features, it was thought there could be only two kinds of galaxies—flat ones with spiral features and ellipsoidal ones. The three-dimensional ellipsoidal shape of elliptical galaxies is hypothesized by astronomers. The reason the three-dimensional shape of elliptical is not known is that all very distant objects appear two-dimensional. To observe the depth dimension of an elliptical galaxy, we would have to be able to observe these objects simultaneously from two separate points in space very far apart (farther apart than the distance from Earth to the next nearest star: Proxima Centauri).

••

Depth Perception

Depth perception is due to the ability to observe simultaneously from two different points in space. The human being has two eyes separated by a small distance (usually about 8–10 cm). This distance difference allows us to perceive depth. The trick is that the view from each eye must have a region of overlap on the other eye's view. If there is sufficient overlap, the brain must reconcile the two different images it is receiving from each eye. This is how humans can tell that one thing is farther away than another, without being able to detect the images overlapping. This principle is used in astronomy to measure distance using the parallax method. This method requires that one observe the same star when the Earth is at two different points in space during its revolution around the Sun. These observations are not simultaneous, but they allow astronomers to measure the distance to stars within about 500 pc of Earth.

••

Looking at the distribution of surface brightness vs. distance from the center of an elliptical galaxy, some elliptical galaxies have light curves that increase, at the center, to a very high value of brightness. (Most elliptical galaxies follow a power law in surface brightness where the surface brightness at any given radius from the center of the galaxy is proportional to the central surface brightness times the fourth root of the radius [$r^{1/4}$].) The unusual "cusp" feature may indicate an over-density of stars near the centers of these galaxies. Kinematically, the cusp is an indicator of chaotic stellar orbits. Such orbits often occur when two galaxies merge. The stars near the center begin to move in chaotic orbits during the merger, and it takes some time for these chaotic stellar orbits to settle into regular orbits.

So-called cuspy light curves, it turns out, are likely to be counter-indicators for supermassive black holes. So, the few elliptical galaxies that have this cusp feature in their surface brightness curves probably do not contain supermassive black holes, but most elliptical galaxies are thought to contain a central supermassive black hole since they lack the cusp feature. Elliptical galaxies as a class, therefore, are thought to contain supermassive black holes in their centers.

Giant Elliptical Galaxies

Giant elliptical galaxies are the oversized version of spheroidal galaxies. These enormous galaxies contain as many as several trillion stars, yet contain almost no gas or dust. The lack of ingredients for star formation is thought to be due to the way in which these behemoth galaxies form.

It is suspected that these galaxies start out as, perhaps, normal-sized elliptical galaxies, at the gravitational center of a dense, rich cluster of galaxies. Then, through a series of gravitational interactions with other members of the cluster, the central galaxy eventually consumes several galaxies from the cluster. Each gravitational interaction causes a burst of star formation that efficiently uses up most of the gas and dust to make new stars. By the time the giant elliptical has formed, almost all of the gas and dust originally contained in the central galaxy and the galaxies it consumed is tied up in stars recently formed.

The gravitational interactions that build up this giant are also the reason the giant retains a spheroidal shape. This process causes the giant to have **triaxial** kinematics. The stars in these galaxies orbit the central supermassive black hole in every plane allowed in three-dimensional space. This motion is called triaxial because each of the three axes that define three-dimensional space are crossed by an orbiting star at some point in its path.

Even more interesting than this bizarre formation process is that giant elliptical galaxies are thought to contain supermassive black holes at their centers. Supermassive central black holes are suspected to exist in these galaxies due to the recent discovery of dusty disks in the centers of these giant galaxies. These dusty disks are apparently disks of material being accreted by the central black hole. The process by which these formed is thought to be the same process already described, cannibalistic gravitational interactions. Galaxies with preexisting central supermassive black holes are consumed by the giant elliptical galaxy, and these central supermassive black holes merge in the center of the giant, creating one or more supermassive black hole(s).

OTHER GALAXIES

Aside from these two major types of galaxies, those with disks and those that are spheroidal in shape, there are other identifiable morphologies. These other galaxies are not nearly as common as the disks and spheroidals, but they are common enough to mention and describe in some detail. For the most part, these other galaxies are some combination of the disk and spheroidal types; however, some of these other galaxies are fundamentally and kinematically different from both disks and spheroidals.

Lenticular Galaxies

The first of these other types of galaxies to be discussed are the lenticular galaxies. These are the "in between" galaxies. They appear to have a weak disk component, yet meet almost all the other criteria to be classified as spheroidal galaxies. When galaxies were first classified by their shapes by Edwin Hubble in the early part of the 20th century, he proposed a possible evolutionary sequence (much like the stellar evolutionary sequence). Hubble thought (quite reasonably) that galaxies must start out very round (as spheroidals) and flatten out over time (becoming disks). In Hubble's scheme, lenticulars were the "missing link" showing that the galaxies could be anywhere in the spectrum of disk to spheroidal.

Hubble's evolutionary sequence has not been shown to be correct or incorrect. The evolution of galaxies is still being pieced together. The timescales for change in galaxies is much longer than that for stars. Understanding morphological evolution, if it exists, will take good modeling and a much better understanding of morphology and how it relates to the physical nature of the galaxy than that which currently exists.

Irregular Galaxies

Irregular galaxies are nothing like either disk galaxies or spheroidal galaxies. For the most part, these are dwarf galaxies. They look a little like the central parts of disk galaxies (the bulges or bars). The most common example of an irregular galaxy is the Large Magellanic Cloud. This is a dwarf galaxy that orbits the Milky Way. It is one of the closest neighbors to our galaxy.

The Large Magellanic Cloud has a bar and a lot of gas and dust (much like a disk galaxy), but no disk (like a spheroidal galaxy). There is a lot of recent and ongoing star formation in this galaxy, and at the same time, there is an underlying old stellar population in this galaxy, indicating that this galaxy has been around for a long time. Kinematically, irregular galaxies tend to look like disk galaxies, in that most of their constituents are moving in two dimensions.

Peculiar Galaxies

This type of galaxy is by far the most unusual. Most galaxies classified as peculiar are actually not really a single galaxy, but a pair (or more) of galaxies in the process of merging into one galaxy. An example of a peculiar galaxy is the antennae galaxy merger pair.

Originally, galaxies given the peculiar designation were so classified by Halton Arp in 1966. Some of these galaxies are pairs, others are called peculiar because they have unusual spectra, but morphologically, they have no distinctive features. Arp's main motivation in creating this new classification was to provide motivation to the astronomical community to investigate further the physical nature of these objects. That some of them have no obvious morphological peculiarities is notable.

RECOMMENDED READINGS

Asimov, Isaac. *Asimov on Astronomy.* New York: Bonanza Books, 1988.

Bennett, Jeffrey D., Megan Donahue, Nicholas Schneider, and Mark Voit. *The Cosmic Perspective.* 5th ed. San Francisco: Benjamin Cummings, 2007.

DeGrasse Tyson, Neil, Charles Tsun-Chu Liu, and Robert Irion. *One Universe: At Home in the Cosmos.* Washington, DC: Joseph Henry Press, 1999.

Elmegreen, Debra Meloy. *Galaxies and Galactic Structure.* Englewood Cliffs, NJ: Prentice Hall, 1997.

Freedman, Roger, and William J. Kaufmann III. *Universe.* 8th ed. New York: W.H. Freeman Company, 2008.

Hawking, Stephen. *On the Shoulders of Giants.* Philadelphia: Running Press, 2002.

Seeds, Michael A. *Astronomy: The Solar System and Beyond.* 5th ed. Pacific Grove, CA: Brooks Cole, 2006.

WEB SITES

Centers for Disease Control. n.d. http://www.atsdr.cdc.gov/substances/toxsubstance. asp?toxid=25

http://www.spitzer.caltech.edu/features/articles/20050627.shtml

The Physics and Relativity FAQ. 1997. http://math.ucr.edu/home/baez/physics/ General/occam.html

6

The Milky Way

Our own galaxy is called the Milky Way. As mentioned earlier, the name for our galaxy in English is related to the Greek myth describing how it came to be. The myth says that the Greek Goddess, Hera, Queen of the Gods, while nursing Heracles spilt her breastmilk in a streak across the sky, creating the Milky Way ("galaxias" in Greek).

In the previous chapter, we discussed the many different kinds of galaxies that astronomers know about. What kind of galaxy is the Milky Way? How do we know what kind of galaxy it is? In the previous chapter, we learned that galaxies are so enormous that humans have yet to travel far enough to leave the Milky Way.

In fact, if you imagine the solar system taking up a football field, our Galaxy would be 50 million times as large—approximately equivalent to 10 times the distance between Earth and the Moon. Interestingly, the furthest humans have traveled in space is to the Moon. On the football field solar system, that distance is equivalent to one centimeter. Humans have hardly explored the solar system, let alone the Milky Way galaxy!

In this chapter we will discuss what is known about the Milky Way in the context of the other galaxies we can study. Even though the Milky Way is our nearest example of a galaxy (just as the Sun is our nearest example of a star), astronomers do not describe other galaxies in comparison to our own as they talk about other stars in comparison to the Sun. Rather, we know less of our galaxy because of our position within it. We cannot get a good view of the whole. It is much like trying to determine the extent of a forest while you are standing within it. So, studies of the Milky Way that have produced interesting new discoveries have occurred, perhaps, more recently than one might expect.

WHERE IN THE GALAXY ARE WE?

In the late 18th century, a famous brother and sister team of astronomers attempted to determine our position in the universe. William Herschel was a musician who became interested in astronomy through books. He enlisted his sister Caroline to assist him in his pursuit of knowledge through astronomy. He started by building telescopes. Shortly after he began observing using the telescopes he built, he discovered the planet Uranus. As a result he was employed as the Court Astronomer by King George III of Britain.

After making several discoveries of comets and natural satellites around the giant planets, William and Caroline concentrated their efforts on attempting to determine the "Condition of the Milky Way." At that time, the concept of the Milky Way was entirely different from what we know today. It was thought that what one could see in the night sky constituted the whole of the universe. The other objects one could see using the most powerful telescopes of the time were thought to be part of the Milky Way. That is, it was not thought that other galaxies existed outside of the Milky Way. Although by this time, astronomers had discovered many objects, now known to be galaxies, at the time they were dubbed simply "nebulae." Astronomers assumed they were clouds of gas that were within the Milky Way, like the Orion Nebula.

So the Herschels went about determining the "condition of the Milky Way" by attempting to map the Milky Way. They made a simple assumption. They assumed that they could see all the stars in the Milky Way. To determine the extent of the Milky Way, one simply had to count the stars one could see. Of course, we now know this is not the case, but in the late 18th century, astronomy was still in its infancy. Photography had not yet been invented, and an excellent astronomer was someone who could draw well what they were seeing through a telescope or someone who could be very meticulous about their measurements of time and position in the sky.

Given the information William and Caroline had about the universe, this was a good assumption to make. They had no reason to believe that stars could have different intrinsic brightnesses or that a star's light might not reach the Earth. So, in their crude attempt to determine the extent of the Milky Way, they found that the galaxy appeared to be irregularly shaped, but slightly elongated along one axis with the solar system located near, but not exactly at the center. From their measurements, the Milky Way extended to the location of what is now known to be the center of our Galaxy on one side and to the outer edge on the other side.

What the Herschels did not take into account was the existence of interstellar dust, which can obstruct the light from a star, hiding it from view to an observer on Earth. Again, it is not that William and Caroline were not smart enough to notice that the universe is dusty. Rather, it is that they had no way of making accurate measurements of how much light was missing due to intervening dust absorbing or scattering a star's light.

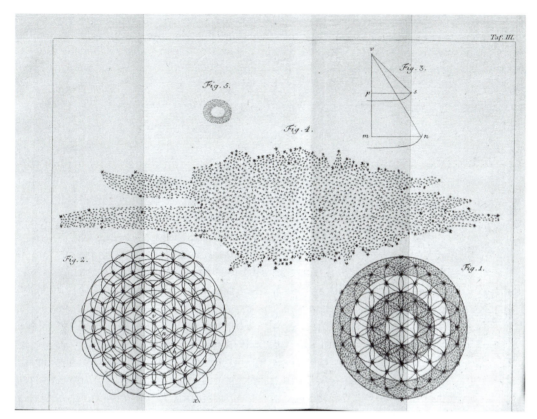

Figure 6.1 This is the map of stars that the Herschels made. The large dot near the middle of the diagram is the location of the Sun. According to what the Herschels were able to measure, it looked like the Sun was located very close to the center of the Milky Way. Shortly after this map was constructed, the Herschels discovered that interstellar dust affected what they were able to see. Once they had accounted for the interstellar dust, their understanding of our position within the Milky Way was modified. [Image courtesy History of Science Collections, University of Oklahoma Libraries.]

Caroline Herschel

Caroline Herschel was born in Hanover, Germany, in 1750 to Issak and Anna Herschel. Issak was a musician (although he did other work to support his family) and hoped that all six of his children would become musicians as well. Caroline was, however, to become her mother's maid. At age 10 she contracted typhus. As a result, her growth was stunted—her adult height was four feet three inches. Her father believed that since Caroline was so small and homely, she would never marry.

At age 22, however, Caroline's elder brother, William, rescued her from Hanover claiming her as his own maid; instead, however, he gave her voice lessons. She became a prominent performer in Bath, England, where she lived with her brother. When William decided to pursue his studies of astronomy and telescope building, he trained his sister in mathematics and telescope building. At the age of 32, Caroline was her brother's apprentice. King George III even gave her a pension of £50 for her work in astronomy.

William's notoriety continued to grow, and Caroline began to do more and more on her own. She detected several nebulae, catalogued them, and discovered eight comets. William married and came to the observatory he had built less and less frequently over time. When William died, Caroline ended her career as an observational astronomer and returned to Hanover, Germany, to live out the rest of her life. Before she died, she catalogued every discovery William and she had made and was proclaimed an honorary member of the Royal Astronomical Society. She was the first woman to ever receive this honor. Caroline Herschel died in 1848 at the age of 98.

••

Over a century later, in 1920, having discovered many nebulae of different shapes and sizes, astronomers began to consider the possibility that the Milky Way was not the extent of the universe. A man named Heber D. Curtis argued that the universe was vast and that the Milky Way was just one small grouping of stars (and gas and dust and planets, etc.) within a universe of many such groupings of stars. On the other side, a man named Harlow Shapley argued that the universe consisted of all that we could see; the nebulae were small, cloud-like objects contained within the Milky Way; the extent of the universe was the extent of the Milky Way. Interestingly, Curtis placed Earth at the center of his universe, whereas Shapley placed Earth relatively far from the center of the Milky Way (his universe).

The Shapley-Curtis debate is an important moment in the history of astronomy. It is at this point in time that our understanding of the universe went from a small area consisting of stars in our neighborhood to the vast expanse that we now understand to be our universe. This change in paradigm is similar to the change every one of us went through when we learned to crawl or walk. Suddenly our universe went from the size of our crib to the size of our home. As a result of this change, there was much more to explore and understand; this new understanding of the universe is what shaped modern astronomical studies.

To settle the dispute, astronomers in the early 20th century began to use the instruments of modern astronomy (photographic plates, spectrographs, and large telescopes) to gather more information about the spiral nebulae whose nature were under dispute. Vesto Slipher of the Lowell Observatory in Flagstaff, Arizona, was the first to point his spectrograph at nebulae and notice that the spectrum was a spectrum of known elements shifted from their original wavelengths. In 1913, Slipher published his first measurement of a velocity for a spiral nebula. It was the velocity of the Andromeda galaxy, which is moving toward us. (Incidentally, this is one of a small fraction of all galaxies known to be moving toward us, as opposed to away from us—this is mostly because the Andromeda galaxy is our nearest large neighbor and is more affected by its gravitational attraction to the Milky Way than to the rate of expansion of space.)

Still, this was not convincing evidence that the spiral nebulae were, in fact, conglomerations of stars located outside the Milky Way. It was not until Edwin Hubble discovered variable stars called Cepheids in the spiral nebulae. A **Cepheid variable** star (see chapter 4) is a star whose intrinsic brightness is related to its period of variability. Since its intrinsic brightness can be known from its period of variability, coupled with the star's apparent brightness, one can calculate the distance to a Cepheid variable. So, by discovering Cepheids within the spiral nebulae, Hubble discovered their distances from us. Hubble's data proved, without any doubt, that the spiral nebulae existed beyond the extent of our Milky Way. Ironically, the astronomer who correctly determined the extent of the Milky Way was Harlow Shapley!

In 1915, Harlow Shapley took on the messy job of determining the extent of the Milky Way. He decided to attempt to correct the mistake the Herschels made in neglecting the effects of dust on starlight. By the early twentieth century, astronomers understood that there was a vast amount of interstellar dust in the universe. This dust made things look redder and even hid some objects from view. Shapley noted that there was significantly less interstellar dust, on average, in the direction of objects known as globular clusters (see chapter 3), which are round-shaped clusters of stars containing as many as a million stars within them. Near the Milky Way, globular clusters tend to be found far from the visible plane of the galaxy, in the halo. There is much less dust in the halo of a spiral galaxy, so these clusters are less affected by dust than other galactic objects.

Shapley decided to use the globular clusters as beacons of light above the fog of interstellar dust. He assumed that the distribution of globular clusters was the same as the distribution of stars within what he thought was the universe. Therefore, if he could just plot the position and distance to each globular cluster, he could create a map of the universe.

Assumptions about the Universe

It is good to reflect, here, on the many mishaps that have occurred because of assumptions humans have made about their position in the universe. When humans thought of the solar system as their universe, many believed that Earth was the center of the solar system. But, with careful observation of phenomena related to the motions of the objects in the solar system, it was determined that Earth could not be at the center of the solar system. Rather, the Sun was at the center of the solar system.

With the assumption that the Herschels made about the distribution of stars in the galaxy, there was no underlying premise that Earth was located at the center, but due to their negligence of the interstellar dust, their result appeared to show that Earth was nearly at the center of the galaxy. Shapley's approach takes dust into account and also has no underlying premise about the location of Earth within the galaxy.

So, the fact that many *expected* Earth to turn out to be at the center of the galaxy was merely a paradigmatic effect. That is, previous studies had led many to believe that Earth must be in a special place in the universe. Nowadays, the paradigm has shifted 180 degrees, so that scientists start out with the notion that Earth *cannot* be in a special place in the universe. This principle is called the Copernican Principle.

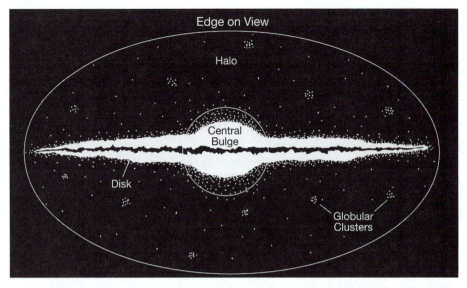

Figure 6.2 This diagram shows the location of the globular clusters in the halo of the Milky Way, far from the interstellar dust located in the disk. Using globular clusters to map out the distribution of the stars in the galaxy avoids the problem of the interstellar dust. It is much like using a light house to navigate around rocks on a foggy ocean. [Credit: NASA, ESA, and The Hubble Heritage Team (STScI/AURA).]

So, Shapley observed many globular clusters and concluded that the location of Earth is *not* in the center of the galaxy. But, this time, unlike in Copernicus's and Galileo's day, the general public was ready to accept the conclusion that the Earth was not at the center of the galaxy.

THE DISK

The most obvious evidence that we live in a disk galaxy is the view we have of our Galaxy. If you observe the Milky Way (best seen in northern hemisphere summer), you will note that from Earth, what we see is a thin band of stars. If we lived in a spheroidal galaxy, it would be impossible to view our Galaxy from an angle that would make it appear as a thin band of stars. Rather, a spheroidal galaxy from the inside would look like a roundish distribution of stars in the night sky. Depending on where were located within a spheroidal galaxy we might see about the same number of stars in every direction, or we might see more stars in one part of the sky (where the center of the Galaxy would be located) and a relatively even distribution of stars in almost every other direction.

To determine how it is that we see a band of stars from within a disk galaxy, picture a CD and imagine yourself somewhere within the disk. The disk of a disk galaxy is thick enough that many stars (perhaps up to 1,000) could exist in the same part of the disk. A good visual analogy that gives an exaggerated thickness to the disk is to imagine that you are a blueberry inside a giant blueberry pancake. Since the blueberry is part of the pancake, the view of the pancake from the perspective of the blueberry would be just the inside of the pancake nearest itself. However, looking carefully, you would notice that there is little or no pancake above or below the blueberry. Since our view of the Milky Way is a thick band of stars (just like the view of the pancake to the blueberry), we must be *inside* the disk of our Galaxy.

The disk of our Galaxy is where the majority of the bright astronomical objects are located. The brightest stars are in the disk of the Galaxy, bright clouds of gas illuminated by bright stars are located in the disk of the Galaxy, and bright dust clouds lit up by stars are located in the disk of the galaxy. Most of what makes up a galaxy is located in the disk of the galaxy.

Open Clusters

Open clusters of stars are groups of stars that are physically related. In an open cluster, all the stars form from the same diffuse nebula of gas and dust. Nearly all the stars form at the same time. So, open clusters contain stars with the same chemical composition that are the same age. The name "open cluster" is given to these stellar associations because they are relatively diffuse objects with no obvious center.

Open clusters are found in the disk of a spiral galaxy. Because of their gravitational interactions with clouds in the disk of a spiral galaxy, open clusters tend to diffuse (lose their members) in a short period of time, compared to the lifetimes of the members. Most open clusters spread their members along their path within a few hundred million years. Open clusters are considered young because the stars of which they are comprised are recently formed. Also many of these kinds of star clusters still contain the dust and gas that was used to form the stars within them; however, it is important to note that the word "young," here, does not mean a few years old; it means, typically, less than 10 billion years old.

Open clusters are characteristically irregular in shape and contain usually only a few tens to several thousand stars. Since open clusters are so diffuse, they are not easy to spot. It usually takes a lot of stellar spectra to determine the members of an open cluster, and even then, there are often members left out or stragglers included accidentally. There are several hundred known open clusters in the Milky Way.

The importance of open clusters in astronomy is discussed in great detail in chapter 3. To summarize, open clusters, along with globular clusters were used to piece together stellar evolution. This was done by putting the members of each cluster (whether open or globular) on a cluster H-R diagram. (H-R Diagrams are discussed in great detail in chapters 2 and 3 of this volume.) Because of the fact that the members of a cluster are the same age (since all stars formed at essentially the same time), it is possible to determine what happens to stars as they evolve.

THE ARMS

The arms of the Milky Way are where there are dense concentrations of gas, dust, young stars, and star-forming regions. Primarily, the arms of a disk galaxy are bright spiral-like features seen within the disk of the galaxy. The arms are bright because the arms are the place where all the ingredients for star formation are present in exactly the right amounts. Also, in the arms, clouds of dust and gas are being pushed together, which causes the gas clouds to break up into small pieces, which collapse further, forming stars. In chapter 7, more will be said about how arms work and what their significance is to the galaxy, dynamically speaking.

Star-Forming Regions

Star-forming regions are places were stars are forming. They are usually clouds of hydrogen gas that are lit up by the young stars forming inside. Most often, these regions are called HII (pronounced "H-two") regions. This is because HII is the astronomical symbol and term for ionized hydrogen. In these star-forming regions the hydrogen is ionized by the intense ultraviolet radiation coming from the most massive stars formed. The radiation from these stars ionizes the gas surrounding these stars out to hundreds, sometimes thousands of parsecs away from the stars themselves. So, what astronomers see most often is a large cloud of ionized gas. Looking more carefully, they may be able to see the actual stars responsible for ionizing the gas they see.

These star-forming regions are important to study. What's happening inside them is a process about which astronomers still speculate. There are, of course, physical models of the process, many of which rely on principles of physics to guide them through the process, but in actuality, astronomers still know very little about the conditions inside these star-forming regions. Even today, simplifications are made to allow theoretical models to be developed. The nature of these regions is still too complicated to unravel.

Astronomers are hard at work trying to unravel the secrets of star formation, nonetheless. It is an important process to understand. Whether star formation has temperature, density, or constituent limitations is still poorly understood. Part of the problem is time. It takes a long time (several million years) for star-forming regions to go from a cloud containing cold, neutral gas to a cloud containing hot, ionized gas and still more time for the cloud to become an open cluster or globular cluster. So, astronomers have to make lots of assumptions about how these processes occur, since there are no data about these regions from millions of years ago and an astronomer can not wait millions of years to answer the question of how stars form.

••

Orion Nebula

The most well-known star-forming region is the Orion Nebula. This nearby, bright star-forming region can be seen with a small telescope. Located in the "dagger" of the constellation Orion, the nebula appears grey-green to the naked eye. However, many beautiful images of the famous nebula can be found online. The Hubble Space Telescope has amassed an amazing image library containing color-enhanced space-based images of the Orion Nebula. The ionized cloud of gas that makes up the Orion Nebula is powered by four extremely bright stars.

The age of the stars in the Orion Nebula is approximately 1–2 million years, making it one of the youngest clusters known in our own Milky Way. The famous nebula is also one of the closest star-forming regions to Earth at a distance of about 500 pc (or 1,500 light-years). As it is one of the closest star-forming regions, it should not be surprising that the Orion Nebula is also one of the smallest known star-forming regions. Since most of the star-forming regions astronomers can study are very far away, they are also very large. If those star-forming regions were *not* large, astronomers on Earth would not be able to see them and study them.

••

Emission Nebulae

Star-forming regions are just one kind of emission nebula. Emission nebulae are clouds of gas that emit light. Usually the clouds of gas are ionized and, therefore, emit radiation due to the recombination of electrons with ions. So, an emission nebula is found anywhere that gas is heated enough to be ionized. Thus, emission nebulae are usually found near energy sources. Most times, the energy source is visible (like a star), but sometimes the energy source is not visible (like a supernova remnant or a black hole).

Star-forming regions are where many stars are forming all at once. In such regions there are usually several massive stars forming. These stars will both produce a lot of light and produce strong "winds" because of the extremely high rates of fusion in their cores. This high rate of fusion can cause instabilities in the star as it evolves. Sometimes, during the protostar phase, these massive stars lose material because the radiation pressure is greater than the pressure of gravity holding the star together, and the star blows apart.

Even in star-forming regions where such massive stars are not forming, the presence of stars amidst gas alone is enough to create an emission nebula. The light from the stars comes in all wavelengths. It takes photons of wavelength 121.2 nm to ionize hydrogen gas (the most common gas in the universe). All stars will radiate these photons since all stars are blackbodies. Stars with higher surface temperatures will radiate more of these photons than stars of lower surface temperatures. So, the extent to which the gas around a star is ionized will depend on the surface temperature of the star. Assuming the gas around a star extends forever, a hotter star will ionize gas to a greater distance than a cooler star.

Another type of emission nebula is a planetary nebula. A planetary nebula is one of the end products of an intermediate-mass star (like our Sun). When this type of star exhausts all the possible methods for fusion in its core, the outer shells of the giant star expand during the last shell fusion phases and never contract back to interact with the core. The shells that expand into space are simply layers of (relatively) dense, hot gas that is ionized and emitting light. The motion of this hot gas can also cause emission from gases with which it interacts as it expands.

Additionally, a supernova remnant is a kind of emission nebula. A supernova is the event that marks the end of the existence of a massive star (one with a main sequence mass greater than about eight times the mass of our Sun). This event is caused by the outer shells of a massive star collapsing back on its iron core. The core collapses further into a neutron star or a black hole, depending on its mass. The shells bounce off of the collapsed core and move outward into space at supersonic speeds. The shells are already hot, ionized gas, and as they move with supersonic speed through the interstellar medium, they collide with neutral gas and ionize it. This collisional excitation (when atoms are excited or ionized because they are involved in a collision with another particle) is distinguishable from photoionization (when an atom is excited or ionized because it interacts with an energetic photon). Collisional excitation causes more low-energy transitions than high-energy transitions, whereas photoionization does the opposite—it produces more high-energy transitions than low-energy transitions.

Still another kind of emission nebula can occur around supermassive black holes. Around these objects, which are found in the cores of some galaxies, material is being accreted onto the black hole by means of an accretion disk. The accretion disk is a thin disk of material that is slowly falling into the black hole. The material in the disk is very dense and very hot. The heat comes from friction and shearing forces acting on the material in the disk. Again, the cause for the emission is not photoionization, but collisional excitation. The atoms are colliding with one another and causing the electrons to either move up to higher energy levels or be removed from the nucleus, ionizing the atom.

Finally, the last type of emission nebula is the neutral hydrogen cloud. As discussed earlier, the neutral hydrogen clouds emits photons because the hydrogen atom has two ground states. One ground state (the more common one) occurs when both the electron and the proton are oriented so that their spins are considered "up." The other ground state occurs when the electron and the proton have opposite spins. Since the proton is a particle that does not easily change its spin, it is the electron that does the flipping. This transition from the common ground state to the ground state of lower energy is commonly known as the spin-flip transition. The photon emitted has a wavelength of 21 cm and is visible in the radio part of the electromagnetic spectrum, so these emission nebulae are only visible in the radio, not the visible part of the electromagnetic spectrum.

Dark Nebulae

A **dark nebula** is one that appears dark against the background. The background could be an emission nebula or other stars. These nebulae appear dark, because they are dominated by dust which blocks out the visible light coming from behind. The most famous dark nebula in the northern hemisphere is the Horsehead Nebula, which is located in the Orion Nebula (a well-known emission nebula located in the dagger part of the constellation Orion). The Horsehead Nebula is difficult to see, even with the aid of a telescope because it is very faint. It is best seen photographed for a long exposure time, so that many photons can be gathered before the image is processed. Our eyes don't work that way—they integrate over a fraction of a second, so they would need a very large telescope to gather more photons in our short integration period to see the nebula. Such large telescopes don't usually have eyepieces for humans to look through—they are filled up with instrumentation for astronomical studies. Against the faint emission nebula, there is a dark feature that humans recognize as having the shape of a horse's head. The dark feature is really a dense cloud of dust.

In the southern hemisphere the most well-known dark nebula is the "Coal Bin." This is an extremely dark region of the sky. It is noticeably darker than any other region not least because it occurs along the line of sight of the Milky Way (which is spectacular in the southern hemisphere) so that it blocks out the light of billions of stars. In comparison to parts of the sky that are considered dark because there are few nearby stars visible, the coal bin is darker than these as well. This dark nebula is also a dense cloud of dust that is blocking the light from the stars behind it.

Reflection Nebulae

Reflection nebulae are found around young stars. Some of the material from which they formed still surrounds the young stars and the dust that is behind the stars, from our line of sight, is reflecting the light from the stars back in our direction. The most famous reflection nebula is the one around the Pleiades. This reflection nebula makes the Pleiades look bluer than they really are. This is because of the properties of the dust. The dust actually scatters the light in such a way that the longer wavelength light continues through the dust, while the shorter wavelength light is scattered so much that it goes back in the direction from which it came. Since blue light is shorter wavelength than red light, the nebula looks blue.

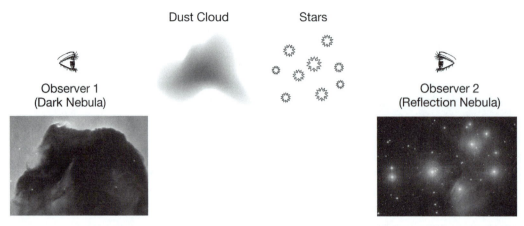

Figure 6.3 This diagram shows how a reflection nebula and a dark nebula are just a dust cloud and a light source (stars) seen from different points of view. If the dust cloud is behind the light source, relative to the observer, the observer sees a reflection nebula where the blue star light is reflected back at the observer. If the dust cloud is in front of the light source relative to the observer, the observer sees a dark nebula where the light from the star is blocked by the dust cloud. [Credit; Diagram: Jeff Dixon; Horsehead Nebula: JPL/NASA, NOAO, ESA and The Hubble Heritage Team, (STScI/AURA); Pleiades: NASA, ESA and AURA/Caltech]

Light Scattering

When light interacts with dust particles in space, it is scattered. Light scattering is the cause of the appearances of many astronomical objects. Dust scatters light so that the longer wavelength photons are scattered the least and continue in the direction in which they were going with little change. Conversely, shorter wavelength photons are scattered the most, so that they go back in the direction from which they came.

This scattering of light is what makes stars seen through dust clouds are not visible or appear much redder than they really are. By the same token, stars in front of dust clouds appear much bluer than they really are because all the blue light they are emitting is being reflected back in our direction by the cloud behind them, causing a blue haze to appear around the stars.

Interestingly, dark nebulae and reflection nebulae are made of the same thing: dust. The dark nebula is how the dust appears when it is in front of the light source, relative to the observer. The reflection nebula is how the dust appears when it is behind the light source, relative to the observer.

THE BULGE

The bulge of the galaxy is what the central part of a disk galaxy is called. In the case of the Milky Way, the shape of the bulge is difficult to determine, since it is not clearly visible from our position within the disk of the Milky

Way. Typically, a bulge contains stars that move triaxially (with orbits crossing through all three planes defining the spatial dimensions).

Towards the end of the 20th century, with the birth and growth of infrared and submillimeter astronomy, it was beginning to seem as though the bulge of the Milky Way might actually be a bar. About two-thirds of the bright spiral galaxies in the universe appear to have a bar feature. This means that most spiral galaxies have a bulge that is elongated, so that it looks like a bar, rather than a bulge. The first evidence for a bar came in the 1980s and 1990s with space-based and ground-based near and far-infrared observations of the central regions of the Milky Way.

It is significant that the evidence came via the infrared regime of the electromagnetic spectrum. The primary reason that infrared detectors were needed is that longer wavelength light can travel greater distances without being disturbed (absorbed or scattered) by intervening media. That is, astronomers see further through the pancake-like disk of the Milky Way when they collect infrared light, rather than visible-wavelength light.

Second, it is important to understand that what astronomers are seeing are very old stars that give off most of their light in the red part of the visible spectrum or the infrared part of the electromagnetic spectrum. So, even if there weren't dust in the way to block the visible light, astronomers might not have been able to see the stars in the center of the galaxy using visible light detectors, because those stars are too faint in the visible part of the spectrum.

The stars in the bulge/bar of the Milky Way are old. Most of the stars in this part of the galaxy are so old that they are no longer main sequence stars, like our Sun. They have evolved off the main sequence and are now red giants. (For more on stellar evolution, see chapter 3.) The most recent evidence has come from the Spitzer Space Telescope (SST). This is an infrared telescope that was launched in 2003 and has been orbiting the Sun, just behind Earth in its own orbit of the Sun. From this vantage point, the SST can see the universe and radio the data it takes back to Earth. Being further from Earth means the SST doesn't need as much coolant as it would if it were in an orbit around Earth. This gives the telescope's detectors a longer lifetime for less weight (a very important consideration for space-based telescopes).

In addition to the many, many stars that make up the bulge/bar of the Milky Way, there is, likely, a black hole in the center of our galaxy. The evidence for the black hole is in the motions of the stars in the center of our galaxy. Stars near the center of the galaxy (in the center of the bulge/bar) are moving far faster than possible to explain using ordinary physics, unless there is a massive object that is not seen in any part of the electromagnetic spectrum. Such an object would have a mass equal to several million times the mass of our Sun and would be much smaller than our solar system. The only object that could meet all three of these criteria is a black hole.

Globular Clusters

Globular clusters are very important indicators of the unseen part of the Milky Way—the halo. Located above and below (as well as in) the disk of the Milky Way, globular clusters are clusters of tens of thousands to millions of stars. The stars are gravitationally bound to one another and form a roundish cluster that looks much like a tiny spheroidal galaxy. Like open clusters, globular clusters probably formed in the disk, where the ingredients for star formation are plentiful. However, unlike open cluster formation, globular cluster formation does not occur anymore. As far as astronomers can determine, no globular clusters have formed in the last few billion years. The youngest known globular cluster is thought to be 5–10 billion years old, but its origin is suspect. It is entirely possible that this globular cluster was not formed in the Milky Way, but was captured by the Milky Way more recently.

The most obvious difference between open and globular clusters is their size. Globular clusters are far more compact than open clusters, but globular clusters contain far more stars than open clusters do. Secondly, their motions are different. The open clusters remain within the disk, moving as other stars in the disk move, whereas the globular clusters inhabit the halo part of the galaxy. Finally, the stars in open clusters contain more metals (heavy elements) than the stars in globular clusters. This difference is thought to be due to the fact that globular clusters formed earlier than open clusters, before the gas from which stars are formed was enriched with metals by the catastrophic ends of the previous generation of stars.

Globular clusters are used in concert with open clusters to piece together stellar evolution. Because stars change so slowly, it is not possible for a human being to observe these changes in a lifetime (or even several lifetimes, for that matter). So, to learn how stars evolve, it was necessary to use clusters. Because all the stars in a cluster (either globular or open) form at approximately the same time and because the stars can be considered to be all at the same distance from Earth, clusters of stars have been very helpful in unraveling this mystery. For more information on the use of clusters in understanding stellar evolution, refer to chapter 3.

RECOMMENDED READINGS

Asimov, Isaac. *Asimov on Astronomy.* New York: Bonanza Books, 1988.

Bennett, Jeffrey D., Megan Donahue, Nicholas Schneider, and Mark Voit. *The Cosmic Perspective.* 5th ed. San Francisco: Benjamin Cummings, 2007.

DeGrasse Tyson, Neil, Charles Tsun-Chu Liu, and Robert Irion. *One Universe: At Home in the Cosmos.* Washington, DC: Joseph Henry Press, 1999.

Elmegreen, Debra Meloy. *Galaxies and Galactic Structure.* Englewood Cliffs, NJ: Prentice Hall, 1997.

Freedman, Roger, and William J. Kaufmann III. *Universe.* 8th ed. New York: W.H. Freeman Company, 2008.

Hawking, Stephen. *On the Shoulders of Giants.* Philadelphia: Running Press, 2002.

Seeds, Michael A. *Astronomy: The Solar System and Beyond.* 5th ed. Pacific Grove, CA: Brooks Cole, 2006.

7

Arms in Disk Galaxies

The most prominent feature of a disk galaxy is its spiral arms. Sometimes these are clearly defined arms, and sometimes the arms are just small spurs appearing as though the disk were made of feathers. The arms are, regardless of their appearance, where star formation is occurring within the disk galaxy. Therefore, the arms are the most interesting part of a disk galaxy.

ARM PATTERNS AND KINEMATICS

The appearance of the arms in a disk galaxy vary considerably. Even if the arms are prominent, the arms of a disk galaxy can have different sizes, winding tightnesses, brightnesses, or lengths from galaxy to galaxy. Symmetry plays a big role in how the arms in a disk galaxy appear.

A galaxy with prominent arms that are clearly attached to the central bulge or bar and can be followed continuously, spiraling outward until they reach the edge of the visible disk is called a **grand design galaxy.** A galaxy with arms that are unstructured and broken into parts so that the disk appears like the fleece of a sheep is called a **flocculent galaxy.** There are actually 12 different arm classifications varying between flocculent and grand design; these classifications include possible causes for the differing appearances of the arms like the presence of a central bar or nearby galaxies.

Figure 7.1 The photograph on the left shows a flocculent galaxy. Notice that the arms are not continuous; rather, they are in pieces or "spurs," giving the disk a "fleecy" look. The photograph on the right shows a grand design galaxy. Notice that the arms are long and continuous from the central part of the galaxy to the outer edge. [Credit: NASA, ESA, S. Beckwith (STScI), and The Hubble Heritage Team (STScI/AURA)]

ARM VS. INTERARM ENVIRONMENTS

In understanding the difference between arm and interarm environments, it is very important to understand that the arms of a galaxy are not a physical entity. Rather, the arms of a galaxy are a region of a galaxy's disk that contains a higher density of gas, dust, and young stars than the rest of the galaxy's disk. But, the important point here is that the stars, dust, and gas that make up the arms of a galaxy are not continuously part of the arms, nor are the arms always made up of the same stars, dust, and gas all the time. Interstellar material is constantly moving into and out of the arms and throughout the disk of a spiral galaxy.

So, the main difference between the environment of an arm and the environment of the disk between the arms (the interarm) is the density of the material, but the material itself is the same both in the arm and between the arms. The only unique feature associated with spiral arms is the existence of newly formed massive stars that lie just at the leading edge of the arms in the disk. The reason for this is that the compression of material that occurs as gas and dust enters the arms causes stars to form. By the time the material has passed through the arm, new stars have formed. Most of the stars will exist for a long time and become part of the disk of the galaxy. The massive stars, however, will not exist for long. In fact, the most massive stars will cease to exist before they can move away from the spiral arm feature, let alone complete a full rotation around the galaxy.

It is for this reason that they are found in great abundance near the place where they are formed, at the leading edges of spiral arms. Not all massive stars are formed in spiral arms, but the spiral arm phenomenon forms stars more efficiently than other processes in galaxies with spiral arm patterns. Since the most efficient star formation occurs in arms, most stars are formed there. So, most massive stars will form in spiral arms, too. As a result, there is a higher density of massive stars on the leading edge of the spiral arms in a spiral galaxy.

As for the other components of the galaxy—gas, dust, lower mass stars—these are distributed almost evenly throughout the disk of a spiral galaxy. Even though these materials are in slightly higher abundance within the spiral arm features of a spiral galaxy, there are gas, dust, and stars everywhere within the disk of a spiral galaxy. So, the difference between the arm and interarm environments is solely the density of the stars gas and dust that make up the disk of the spiral galaxy.

ARM CLASSES

As mentioned earlier, there are 12 different arm classes that describe the structure of the spiral arms in a spiral galaxy. Most of the arm classes describe the presence (or lack thereof) of spiral arms, the number of arms, the extent of the arms (from center to outer visible edge), and the appearance of the arms, but some include information about nearby galaxies and the presence of a central bar, rather than a bulge. Arm classes 1–4 are galaxies that are classified as flocculent galaxies due to their fleecy, disorganized appearance. Arm classes 5–12 are galaxies that are classified as grand design galaxies due to their long, prominent, symmetric arm structure. The appearance of these galaxies tells the story of the underlying dynamics. While most things in astronomy are not as they first seem, in this case, the shape, length, prominence, and symmetry of the arms of spiral galaxies reveal something about the physical conditions within that galaxy's disk.

DENSITY WAVES

As discussed at length above, the spiral arms of a galaxy are not physical entities, but rather define a place within the disk where there is an over-density of material. The material moves through this place creating new stars, which makes this place seem as if there are more stars there than anywhere else in the galaxy; however, there are just more *new* stars there than anywhere else in the galaxy. Interestingly, though, the place is not stationery, either. The material moves through the over-dense region and the over-dense region moves, too, though usually much more slowly than the material moving through the region. This movement of a region of over-density is described as a density wave.

The density wave theory is the explanation for spiral arms discussed in this chapter. It further includes the fact that these waves are a result of gravitational interactions within the disk combined with the underlying gravitational potential well of the galaxy. The theory behind where the waves come from is more tenuous. There is some evidence that density waves can be caused by resonances in the orbits of stars in the disk of the galaxy, and there is also some evidence that density waves can be caused by a gravitational perturbance (like an interacting neighbor or a bar).

The Winding Dilemma

The winding dilemma is the famous explanation for why density waves must be the correct mechanism for spiral arms. Because a large fraction of disk galaxies have spiral arms, we can conclude that spiral arms are long-lived features. If they were not long-lived, we would expect that finding a galaxy with prominent spiral arms would be difficult to do (i.e., spiral arms would be rare). Since they are common, we conclude that they must be around for a long time.

So, the idea that these features might be physical becomes a problem. We know, from observing the spectra of gas in galaxies (and even some stars in galaxies), that the gas and stars move around the galaxy at a nearly constant speed. In other words, stars near the center are moving with the same speed as stars near the outer edge. If spiral arms were a solid, physical part of the galaxy containing the same stars all the time, the spiral arms would wind up.

In the figure below, there are two examples of how material in a galaxy could move. The example on the left shows solid-body rotation. In this case, like a merry-go-round or Frisbee, the material near the center of the galaxy moves so that the material near the outer edge is always on the same line.

The Winding Dilemma

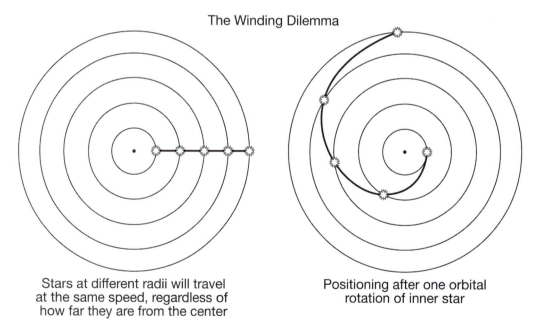

Stars at different radii will travel at the same speed, regardless of how far they are from the center

Positioning after one orbital rotation of inner star

Figure 7.2 The two drawings demonstrate why density waves are necessary to explain spiral arm patterns in galaxy disks. They also show the winding dilemma. Since the stars and gas in galaxies rotate around the galaxy at the same speed, no matter how far they are from the center, this means stars that were once in the arms would eventually move out of them. [Jeff Dixon]

This means that in equal intervals of time, the material on the outer edge will go farther than the material near the center (like the second hand on an analog watch). If galaxies rotated this way, the arms could be explained as solid or physical features containing the same material all the time. However, we know that galaxies don't move that way.

Instead, galaxies move like the example on the right (called differential rotation), so that in equal intervals of time, material near the outer edge of the galaxy moves the same distance as material near the center of the galaxy. This means that in less than one rotation of the material near the center of the galaxy, the material on the outer edge has lagged behind. Take this forward a few more rotations and your spiral arms will wind up, wrapping so tightly that they can no longer be identified as spiral arms.

The rotation period of a galaxy is only a few hundred thousand years—less than the length of existence of most stars. If the above scenario were true, then on very short timescales spiral arms would disappear. Yet, we see spiral arm patterns in many disk galaxies today. This is the winding dilemma. Using this model of differential rotation, which is based on observed galaxy motions, it is clear that spiral arms cannot be explained as a solid entity because the spiral features would disappear very quickly (within a few rotations) and we would not expect there to be any spiral arm features visible in the universe today (or at least, it would be very rare).

Spiral Density Waves

The theory of spiral density waves must account for both flocculent and grand design galaxies. For that reason, simply constructing a model for how they work in spiral galaxies is not sufficient—all mechanisms that cause spiral arms must be nonfunctional in flocculent galaxies to pass muster. Further, the surviving theory must explain all the varieties of spiral arms found (arm classes 5–12), which run the gamut from a single arm to more than two arms and from broad, fluffy arms to narrow, crisp-looking arms.

The current state of the **spiral density wave** theory is that a combination of mechanisms probably work together to create the variety we see in spiral structures. The grand design galaxies are most clearly explained using density waves without the need for an instability to drive the density waves. More intermediate structures require a perturbation or instability within the disk to explain the observed structure. Flocculents seem to be best explained by stochastic star formation theories (not using spiral density waves at all).

Stochastic Star Formation

Stochastic star formation is the mechanism that produces flocculent galaxy structure. This structure is found in disk galaxies that lack spiral arm structures. Instead of clear spiral arms, flocculent galaxies have loose, fleecy-looking spurs of star formation that are disconnected from one another. Stochastic star formation is the only proposed mechanism that successfully explains and predicts flocculent galaxies.

In stochastic star formation, each event of star formation is involved in a domino effect process such that it contributes to the next event of star formation. As far as astronomers can determine, star formation is triggered by some gravitational instability in a cloud. A gravitational instability can be caused by many things. The process of stochastic star formation identifies the cause of such instabilities as supernova explosions and stellar winds (both from young, massive stars and from T Tauri stars).

During the process of star formation massive stars go through phases that cause them to have extremely strong stellar winds. This is part of the massive star's journey to hydrostatic equilibrium (when the star is balanced by the forces pushing it inward and those pushing it outward). During the massive star's journey, the outward forces will sometimes overpower the inward forces, causing massive ejections of stellar material at extremely high speeds. These mass ejections can trigger star formation if they reach a cloud containing the right materials (enough molecular and neutral hydrogen) and cause a gravitational instability.

Less massive stars, like the Sun, also undergo periods of instability that cause strong stellar winds. This occurs during the T Tauri phase of stellar evolution. During the T Tauri phase, stars eject material. This occurs because they are accreting material from a protoplanetary disk and some of the material being accreted is ejected in the form of jets (usually perpendicular to the accretion disk). For more information about these stars, see chapter 4. The jets from T Tauri stars can cause star formation to occur if their jets reach a cloud containing the right ingredients and cause a gravitational instability.

In addition to these mechanisms, supernova events are massive explosions of high mass stars. These explosions cause material to move extremely fast in all directions from the point of origin. Since massive stars can only be formed in large star-forming complexes where there is enough material to produce such stars, supernova events will occur in those same locations, for the most part. (Some exceptions include binary systems where mass transfer can occur as well as Type I supernovae.) Supernova explosions were long thought to be the only possible mechanisms for starting star formation. These explosions move material at very high speeds. Such fast-moving material can start star formation if it interacts with a cloud containing the right materials and cause a gravitational instability.

ROTATION CURVES OF GALAXIES

In the above discussion about the winding dilemma, it was briefly mentioned that galaxies rotate differently than a solid disk (like a merry-go-round or a Frisbee). One might think, alternatively, that since a galaxy is made up of stars and gas and dust, that the objects within it might just follow **Keplerian motion,** like the planets in our solar system. That, too, would be incorrect. Instead, galaxies rotate *differentially*. This means that different parts of the galaxy take different amounts of time to complete one full rotation.

This realization put into question many things that astronomers thought they could simply assume about the rest of the universe. For example, it put into question whether gravity works the same in all parts of the universe. One may be surprised to know that this event of deep questioning of our foundational understanding of the universe occurred in the 1970s, in modern times, when our understanding that the universe was vast and contained billions of galaxies had existed for almost half a century. It all came as a result of the work of one woman, who had simply tried to measure the rotational speeds of spiral galaxies.

MEASURING ROTATIONAL SPEED

Measuring rotational speeds of spiral galaxies is a fairly straightforward process. It is done most easily by measuring the rotational speed of neutral hydrogen gas. In the advent of radio astronomy in the early 1960s, it was quickly discovered that neutral hydrogen gas emits radiation at the wavelength of 21 cm. Hydrogen atoms emit this long wavelength, low-energy radiation due to an extremely rare transition in the hydrogen atom that is not usually observed on Earth.

Using the Bohr model of an atom, we know that hydrogen has many energy levels. The lowest energy level, level 1, is called the "ground state." This energy level is determined by the position, or orbital, of the electron. It turns out that this is not, actually, the lowest energy state of a hydrogen atom. The reason is that electrons and protons have a property called "spin." For these particles, spin can either be pointed up or down. When both the electron and the proton have the same spin (both up or both down), the energy of the atom is higher than if the electron and the proton have opposite spins (one up and one down). In the case of neutral hydrogen, the electron is the particle that can flip. This usually occurs about once every 10 million (10^7) years. Since giant clouds of neutral atomic hydrogen contain as many as 10^{65} atoms, the probability of seeing emission due to this spin-flip transition at any given time is very high.

So, neutral hydrogen clouds of gas can be readily detected using the emission from the spin-flip transition that is detectable in the 21 cm band of the radio regime of the electromagnetic spectrum. This means that one can measure the radial motion of neutral hydrogen gas using the principle of the Doppler Effect. The Doppler Effect is the lengthening or compressing of a wave due to the relative motion of its source and an observer.

Doppler Effect

The Doppler Effect is an effect on waves due to motion. The most commonly known example of the Doppler Effect is the sound a train makes as it passes you. (Actually, it works with any sound source that is moving relative to you—an airplane, a race car, an emergency vehicle, a car with a loud radio, etc.) In this example, as the source approaches the listener (or the listener approaches the source), the sound waves are received in a compressed form. That is, the listener hears a higher pitch. As the source recedes from the listener (or as the listener recedes from the source), the sound waves received are in an extended form (stretched out). That is, the listener hears a lower pitch.

Interestingly, Doppler presented this idea in the context of both light and sound waves as a proposed explanation for why stars in binary systems have different colors. He proposed that their relative speeds would cause a shift in the light we received making them change color with their motion. He was incorrect that this change in color would be great enough to perceive, but he was correct in that this change in color does occur.

By measuring the amount of shift in the 21 cm line observed, astronomers can determine how fast a cloud of neutral hydrogen is moving toward or away from Earth. Because the shift indicates radial velocities only (motion toward or away from the observer), it is only useful for measuring the rotational speeds of disks that are nearly edge-on, as viewed from Earth. Light that is shifted to shorter wavelength is indicative that the source and the observer are getting nearer to one another. Light that is shifted to a longer wavelength is indicative that the source and the observer are getting farther apart.

Because stars, gas, and dust are all in the disks of spiral galaxies and because ionized gas moves with the disk just as much as neutral gas does, rotation curves of galaxies can also be measured using emission lines of ionized hydrogen (called HII lines). In particular the H-alpha line (the first line in the Balmer series of emission lines of hydrogen, indicating an energy state transition from orbit three to orbit two) can be used to determine the rotation curve of a galaxy. This emission line is in the visible part of the spectrum, so it is readily observable and simple spectrographs can record such observations. When observing galaxies that are nearly edge-on, (or even only slightly edge-on), the shift in light received is measurable because the speeds are very large. The Doppler shift equation that relates wavelength (λ) to speed (v) is shown here:

$$v = \frac{c\,\lambda_o - \lambda_e}{\lambda_e}$$

where v is the speed of the source relative to the observer, c is the speed of light, λ_e is the wavelength of the light emitted by the source, and λ_o is the wavelength of the light observed by the observer. If the wavelength observed is longer than the wavelength emitted, the light is being shifted to longer wavelengths (towards the red end of the visible part of the electromagnetic spectrum), meaning the distance between the source and observer is increasing (either the source is moving away or the observer is). If the wavelength observed is shorter than the wavelength emitted, the light is being shifted to a shorter wavelength (towards the blue end of the visible part of the electromagnetic spectrum), meaning the distance between the source and the observer is decreasing (either the source is moving towards the observer or the observer is moving towards the source).

EXPECTED ROTATION CURVE FROM THEORY

As briefly mentioned in the beginning of this section of the chapter, astronomers expected that, since galaxies are not solid bodies, they would exhibit something not unlike the Keplerian motion observed in the solar system. In our solar system, the planets (along with other solar system bodies) orbit a central mass (the Sun) due to the fact that they are gravitationally bound to that mass and to one another. Since the galaxy is made up of objects that are gravitationally bound to one another, the motion of the galaxy should be governed by the same principles as the motion of the solar system.

In addition, astronomers considered the mass distributions of the solar system, the galaxy, and a solid body. A solid body, like a merry-go-round or Frisbee, moves so that all parts complete a rotation at the same time. The

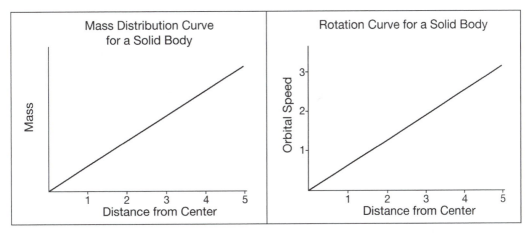

Figure 7.3 The graph on the left shows a mass distribution curve for the solar system. Notice that the mass of the material within a certain radius decreases exponentially as the distance from the Sun increases. This means that as you go farther out from the center, the amount of mass decreases. The graph on the right shows a rotation curve for the solar system. Notice that the material closer to the center moves more quickly than the material farther from the center. Again, the decrease in rotation rate is exponential with increasing radius, mimicking the mass distribution. [Jeff Dixon]

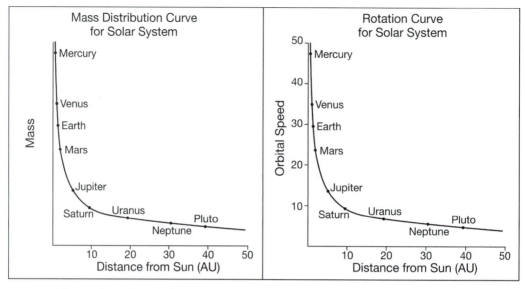

Figure 7.4 The graph on the left shows a mass distribution curve for the solar system. Notice that the mass of the material within a certain radius decreases exponentially as the distance from the Sun increases. This means that as you go farther out from the center, the amount of mass decreases. The graph on the right shows a rotation curve for the solar system. Notice that the material closer to the center moves more quickly than the material farther from the center. Again, the decrease in rotation rate is exponential with increasing radius, mimicking the mass distribution. [Jeff Dixon]

mass distribution in such an object is fairly uniform. That is, there is the same amount of mass (approximately) in 1 cm^2 at the center of a Frisbee as there is in 1 cm^2 at the edge of a Frisbee.

The solar system moves with Keplerian motion. This means that the bodies in the solar system follow Kepler's laws of planetary motion, which, loosely, state that objects that are farther from the center (Sun) move slower and objects that are closer to the center move faster. The mass distribution of the solar system is similar to the rotation curve. Since most of the mass of the solar system (99.8%) is contained within the central body (the Sun), the mass distribution shows a rapid decrease with increasing distance from the Sun.

It should be clear at this point that the mass distribution of the rotating system gives some hint about the rotation curve of the system. In both cases examined, the rotation curve and the mass distribution curves match up very well. So, the rotation curve of a galaxy could be used to determine the mass and mass distribution of that galaxy.

Galaxies are more like the solar system than like a solid body, since they are made up of many parts that are all moving around a gravitational potential well (the galactic center). Astronomers did not know the mass distribution of a galaxy, but it could be surmised by considering the light curve of the galaxy. If one makes the assumption that everything that contributes to the mass of the galaxy emits light, then one can observe the light and map its intensity

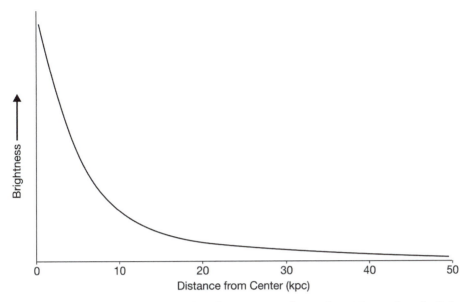

Figure 7.5 The graph shows a light distribution curve for a galaxy. Notice that the light decreases exponentially as the distance from the center increases. This means that as you go farther out from the center, the mass decreases exponentially. [Jeff Dixon]

in relation to the distance from the center of the galaxy to determine the mass distribution of the galaxy. Since all the matter we know to be contained within galaxies does emit light, this seems like a good assumption to make.

Looking at the light distribution of any spiral galaxy (see example below), it is easy to see that the implied mass distribution would indicate that the galaxy rotation curve should look a lot like the rotation curve of the solar system.

OBSERVED ROTATION CURVES

What was observed was completely unexpected and did not match either of the two models considered. When it was observed in a few galaxies, astronomers thought it might be a strange anomaly, but when it appeared to be pervasive and most, rather than some, galaxies exhibited this behavior, it called into question what we thought we knew about the universe—namely the law of universal gravitation.

So, what did astronomers find? They found that galaxies do not rotate like solid bodies, nor do they follow a Keplerian rotation curve. Rather, galaxies rotate *differentially*. This means that the stars, gas, and dust in the galaxy rotate so that the amount of time it takes for material at any given radius to complete one full rotation is different than at any other radius.

What does it mean that the rotation curve does not match the predicted mass distribution? Well, it means one of two things: either the mass distribution is wrong (i.e., light does not trace mass), or the rotation curve

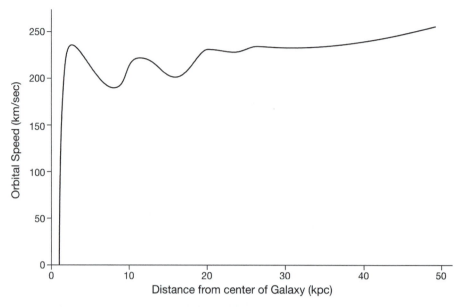

Figure 7.6 The graph shows a rotation curve for a galaxy. Notice that, while there is a steep rise from the center of the galaxy to a point that marks the edge of the bulge (or central) region, the material within the disk closer to the center moves at approximately the same rate as the material in the disk farther from the center. [Jeff Dixon]

indicates that some force, other than gravity, is working to keep the galaxy together.

IS NEWTON WRONG, OR DOES DARK MATTER EXIST?

The rotation curves of galaxies puts into question the assumption that Newtonian mechanics can explain the motions of everything in the universe. Astronomy and the study of the universe starts with the basic assumption that the forces governing any observed motion follow Newton's laws. Astronomers cannot go to different parts of the universe and measure the gravitational constant to make sure it is the same everywhere in the universe, they have to work on the assumption that it *is* the same everywhere in the universe.

If all the matter in the universe is visible, then the distribution of mass of a galaxy should look like its rotation curve, but they are almost opposite one another. This cannot be explained by Newtonian mechanics, but no other set of laws of motion seem to be able to explain what is observed. Either we can use Newtonian mechanics to calculate a mass distribution based on the rotation curve observed, or we assume that the mass distribution is the same as the rotation curve speed distribution and throw out Newtonian mechanics in this case.

Astronomers decided that Newtonian mechanics was too helpful to throw away, but that the data were too convincing to ignore. The compromise is that the mass distribution is not what astronomers expected. In fact, the rotation curves indicate that there is matter in galaxies that does not emit any light. This matter is now called "dark matter."

PROPOSED FORMS OF DARK MATTER

So, the proposal is that the data are correct and that Newton's laws still hold. This implies that there is matter that does not radiate. This matter must be very strange matter. It cannot simply be normal matter, or can it?

Over the years there have been many proposals for the composition of dark matter that explain the observations. The two most long-lived candidates are massive compact halo objects (MACHOs) and weakly interacting massive particles (WIMPs).

Massive Compact Halo Objects

MACHOs can be anything from black dwarfs (the cold, dead remnant cores of intermediate-mass stars like our Sun) to black holes (the remnants of massive stars). Such massive compact objects would not radiate any electromagnetic radiation and, probably, exist in fairly large numbers in the halo (spheroidal component) of galaxies like the Milky Way. These objects should exist in the halos of spiral galaxies because the halo is one of the oldest parts of a spiral galaxy, so the stars that are there are ones that formed near the beginning of the time that the galaxy itself formed. Stars of a wide range of masses would have formed at that time, and the remnants of those that have already ceased to exist would be black holes and black dwarfs.

In the last few decades, astronomers have found evidence to support the hypothesis that MACHOs exist. A group of astronomers have been looking for evidence of MACHOs by looking for secondary evidence that they exist. Since MACHOs, by their definition, do not radiate any electromagnetic radiation, the only way to detect a MACHO is to observe its interaction with an object in the halo that does radiate electromagnetic radiation. The clever experiment dreamed up to test this hypothesis was to look for microlensing events.

Microlensing is the process of a massive, dark object altering the shape of space-time and causing light from a more distant object to be amplified. This is a phenomenon of gravitational lensing.

··

Gravitational Lensing

Gravitational lensing is a phenomenon predicted by Einstein in his general theory of relativity. In his theory, he depicted the universe as a four-dimensional entity that has three dimensions of space and one dimension of time. To visualize this, think of a sheet that can stretch. The sheet represents space-time. A massive object, like the Sun, would be depicted as a bowling ball, for example, on this stretchable sheet. The mass of the Sun would stretch the sheet. In this way, Einstein talked about massive objects stretching space-time. Since everything, including light, can only travel along the sheet, the presence of the massive object changes the way the light will move.

We can observe the change in the shape of space-time by noting that the planets in our solar system are stuck orbiting our Sun because they cannot get out of the dip in the sheet of space-time that our Sun has created (sometimes called a gravitational potential well). What our understanding of space-time did not predict, but Einstein's did, was the bending of light due to the presence of the Sun. Einstein's understanding of the universe predicted that the light from the stars behind the Sun would be bent as they passed around the Sun to be viewed on Earth. So, during an eclipse of the Sun, astronomers carefully measured the positions of the stars nearest the Sun and found that Einstein was correct. The light from these stars did, indeed, appear to have been bent in exactly the way Einstein predicted. (The stars were not in exactly the place they should have been, had their light passed by the Sun without being bent.) Further, it is possible to see the light of distant galaxies bent around more nearby galaxies all over the universe.

The effect of a **gravitational lens** increases the amount of light that reaches the observer. It does not change the color of the light or the distribution of the light across the spectrum, so one would still be able to accurately measure a red shift and a blackbody temperature from the light of a gravitationally lensed object. What makes this brightening of light distinguishable from the brightening observed in a variable star is the smoothness of the light curve and the sharpness of the peak intensity.

• •

MACHOs have actually been detected through microlensing phenomena. Since the late 1990s, observers have been detecting these microlensing events and identifying MACHOs. The problem is that there just aren't enough MACHOs, nor do we know enough about their physical nature, to account for the rotation curves of spiral galaxies.

Weakly Interacting Massive Particles

WIMPs is another way to explain the flat rotation curves of spiral galaxies. Perhaps, particles like neutrinos have a mass that is large enough, that when added up, accounts for the mass distribution indicated by the light curve. We know, for example, that neutrinos are created by the billions within the core of every star in the Galaxy and that they travel for great distances without interacting with anything. Further, billions of neutrinos are also created during supernova events. So, spiral galaxies, like the Milky Way should be full of neutrinos.

Neutrinos are extremely low-mass particles. The first detection of a neutrino was in 1953. This detection was of an electron neutrino. There are three kinds of neutrinos, each one is named after the lepton particle associated with it. There are muon neutrinos and tau neutrinos. The tau neutrinos are the most massive, followed by the muon neutrinos, and then the electron neutrinos.

The electron neutrino is produced through a weak interaction known as beta decay. This interaction occurs in the cores of stars as part of the fusion process. In the first step of the proton-proton chain, when two protons collide and form a deuterium atom (which is made of one proton and one neutron), a neutrino is produced. The changing of one of the protons into a neutron is a weak interaction. It is a form of the beta decay reaction.

In beta decay, a neutron decays into a proton, an electron, and an electron neutrino,

$$n \rightarrow p + e^- + v_e$$

where n is a neutron, p is a proton, e^- is an electron, and v_e is an anti-electron neutrino.

Just switching the positions of the proton and the neutron in the reaction, we can see the nuclear reaction that occurs in stars:

$$p \rightarrow n + e^+ + v_e$$

In this reaction, due to conservation of charge, the electron must be an anti-electron (or positron), and due to conservation of matter, the anti-electron neutrino becomes an electron neutrino.

As a result, we can see that in stars like the Sun, electron neutrinos should be produced in extremely large numbers. Astronomers are very interested in detecting the solar neutrinos because it would be more evidence to support the models of how stars produce energy. However, the solar neutrinos, so far, still present a problem. Neutrino detectors have not been able to detect the expected amount of solar electron neutrinos. There are some ideas as to why scientists have, as yet, been unable to detect the expected amount of neutrinos. One idea is that neutrinos have the ability to change type when they interact with matter. That is, an electron neutrino can become a muon neutrino or a tau neutrino. This could explain why the number of electron neutrinos is lower than expected, but there is not yet evidence that this phenomenon can occur, so the solar neutrino problem is still a problem.

Nonetheless, it is true that our solar system is full of neutrinos. Some come from our Sun, some are cosmological in origin. There are about 10 neutrino detectors located on Earth. Each one uses a slightly different method to detect neutrinos and calculate how many actually passed through.

....

Neutrino Detectors

Neutrino detectors are many and varied in their methods to detect the elusive neutrino. Most are scintillation detectors, which means they detect something called "**Cerenkov radiation,**" which is produced when neutrinos interact with other particles. This radiation is due to the fact that the particles produced move faster than the speed of light through the scintillating medium.

The Super Kamiokande is one of the most well-known neutrino detectors located 1,000 meters below ground in the Mozumi Mine in the Kamioka area of Hida, Japan. This detector is simply a huge tank (50,000 tons) of ultra-pure water Surrounding the water are over 11,000 photomultiplier tubes that can detect the very faint light of Cerenkov radiation that is produced when neutrinos react with the water.

A similar detector, known as the Sudbury Neutrino Observatory, located about 2,000 meters underground in the Creighton Mine in Sudbury, Ontario, Canada, uses what is known as "heavy" water as its medium. Heavy water is just water where the hydrogen atoms are replaced with deuterium atoms. Since deuterium is just hydrogen with an extra neutron, the water molecules are more massive than water made with the more common isotope of hydrogen; hence, the name heavy water. Using heavy water in the neutrino detector has the advantage of allowing the detection of muon and tau neutrinos, instead of just electron neutrinos.

Another type of detector, like the one near Leads, South Dakota, contain a large amount (470 metric tons) of chlorine-containing fluid, like tetrachloroethylene. When a neutrino interacts with the chlorine it changes it into argon. The argon can be removed and used to measure the number of neutrino detections. The detector in Leads was the first to measure the solar neutrino deficit. A similar design uses gallium, which is turned into germanium when it interacts with neutrinos.

So there are billions of neutrinos coming out of stars, like our Sun. Still, the neutrino mass is so small, even when we add up billions of billions of them, we still don't get enough mass to explain the flat rotation curves of spiral galaxies.

At this point in time, astronomers agree that it is likely that some combination of MACHOs and WIMPs must account for the flat rotation curves of spiral galaxies. The idea of searching for other types of matter (like axions) has been passed over at this point in time. Astronomers seem relatively satisfied with MACHOs and WIMPs as the majority contributors to the dark matter.

RECOMMENDED READINGS

Asimov, Isaac. *Asimov on Astronomy.* New York: Bonanza Books, 1988.

Bennett, Jeffrey D., Megan Donahue, Nicholas Schneider, and Mark Voit. *The Cosmic Perspective.* 5th ed. San Francisco: Benjamin Cummings, 2007.

DeGrasse Tyson, Neil, Charles Tsun-Chu Liu, and Robert Irion. *One Universe: At Home in the Cosmos.* Washington, DC: Joseph Henry Press, 1999.

Elmegreen, Debra Meloy. *Galaxies and Galactic Structure.* Englewood Cliffs, NJ: Prentice Hall, 1997.

Freedman, Roger, and William J. Kaufmann III. *Universe.* 8th ed. New York: W.H. Freeman Company, 2008.

Hawking, Stephen. *On the Shoulders of Giants.* Philadelphia: Running Press, 2002.

Rubin, Vera C. *Bright Galaxies, Dark Matters.* New York: AIP Press, 1996.

Seeds, Michael A. *Astronomy: The Solar System and Beyond.* 5th ed. Pacific Grove, CA: Brooks Cole, 2006.

8

Galaxy Interactions

In this chapter we will discuss the interactions of galaxies. Although the average separations of galaxies in the universe are great, at times galaxies do interact. This has occurred in the distant past, when the universe was not as vast as it currently is. It also has occurred in the recent past, due to galaxies just being close enough to one another to gravitationally interact.

Galaxy interactions can be mild (just a distant gravitational interaction that keeps two or more galaxies closer together than typically expected) or violent (such as a merger between two galaxies that results in both using up all the star forming materials to make stars and losing their original shapes). The type of interaction between two galaxies depends a lot on the environment and the initial conditions of the interaction (like speed, angle, and mass). In this chapter, we will run the gamut of galaxy interactions and discuss all the possibilities and outcomes.

MERGERS AND CLOSE ENCOUNTERS

Mergers and close encounters are the most violent interactions between galaxies. A merger is when two galaxies interact gravitationally and only one galaxy is left at the end. In this scenario, the two galaxies that interacted have merged to become one, new galaxy. A close encounter, on the other hand, is when two galaxies interact and get very close to one another, but remain two separate galaxies. A merger may be preceded by several close encounters.

Galaxy interactions occur on enormously long timescales. No human has ever witnessed a galaxy interaction in its entirety. We see snapshots of galaxy interactions in various stages and can surmise the conditions of the interaction using models. In extragalactic astronomy (the study of galaxies other than the Milky Way), models that simulate galaxy interactions are widely used. Generally, astronomers try to mimic the snapshots they observe in the universe by making reasonable estimates of the initial conditions of the interaction and creating a model that simulates that interaction. If the model can create something similar to what is observed, then astronomers can gain some insight into how these interactions may occur.

Astronomers use computers to create models of galaxy interactions. The models are actually simulations of galaxies. Typically, a galaxy model will consist of millions of particles that interact with one another in the same way stars would interact with one another, as point masses through gravitational interactions. Stars behave as "collisionless" particles; that is, stars do not collide often. It can happen, but it is rare because the mass of a star is so concentrated.

As you know, galaxies contain billions of stars and at least as much mass in the form of gas, which does not interact the same way stars do. Because the gas clouds in galaxies are so spread out, they do collide. In fact, that's the primary mechanism for star formation, through cloud collisions. Since the models do not contain gas particles and only millions of star particles, these models can only approximate an actual galaxy collision; however, even these primitive models do a very good job of imitating nature.

It may seem odd, but this process is easiest to model when the galaxy interactions are more nearly catastrophic. With weaker galaxy interactions, it is more difficult to limit the possible initial conditions using modeling. It appears that the models scientists use are able to very consistently reproduce extreme galaxy interactions that result in unusual formations. More modern models, which take into account what astronomers have learned about galaxies in recent years, are more authentic than earlier models. For example, now astronomers know that dark matter halos are very common and usually massive, not to mention often extend beyond the visible disk of the galaxy. This fact was not incorporated into early models of galaxy interactions. Now that the models can incorporate this feature, they do a much better job of mimicking what is seen in nature.

Additionally, the inclusion of gas particles in more recent versions of these models has made a significant difference in how well they mimic what is observed. Originally, gas was not included because it was too time-consuming to include it in the programming. Now that computers have faster processors, including gas in galaxy models is feasible. As a result the models are much closer to mimicking nature than before.

COMPACT GROUPS OF GALAXIES

Galaxies can be in pairs, groups, or clusters. A pair of galaxies would include exactly two galaxies. A group or a cluster of galaxies is more than two galaxies. The difference between a group and a cluster has more to do with the density

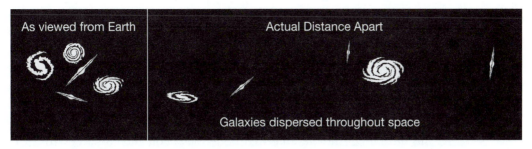

Figure 8.1 This diagram shows how the chance alignment of two galaxies can make it appear that they are interacting, when, in fact, they are hundreds of kpc apart. Some compact groups have chance alignments, rather than physical interactions. Others may contain both chance alignments and physical interactions, while still others are entirely physically interacting. [Jeff Dixon]

of galaxies than the number of galaxies. Clusters have more galaxies per unit volume than groups do. Compact groups have a small number of galaxies (usually less than 10) that appear to be so close to one another that their densities (number of galaxies per unit volume) approach those of clusters.

Compact groups are interesting to astronomers because some of the "snap-shots" astronomers have that show galaxies interacting with one another are from compact groups of galaxies. While many compact groups of galaxies are physical groups that are gravitationally interacting with one another, some are just chance alignments. That is to say that some of the galaxies in compact groups are not really as close together as they appear to be. Instead they may be very far apart in radial distance from Earth.

Galaxies are typically about 20–30 kiloparsecs (kpc) in diameter, and are typically separated by about 1 megaparsec (the equivalent of 1,000 kpc). In compact groups, galaxies appear to be separated by less than a galaxy diameter; chance alignment, however, can make galaxies appear to be much closer together than they actually are. Galaxies could be separated by as much as 100 kpc in radial distance, but because their separation is along our line of sight, they look like they are separated by less than one galaxy radius.

The compact groups that are physically interacting are good laboratories for understanding galaxy interactions. The galaxies in these compact groups are very close together and are gravitationally bound to one another. Because the galaxies are so close together, their interactions are violent and produce interesting features that allow astronomers to test some of their ideas about galaxy interactions. It stands to reason that compact groups are the most likely sites of galaxy interactions.

PAIRS OF GALAXIES

The simplest case of galaxies interacting is with pairs of galaxies. While the most interesting interacting galaxies are in compact groups, the most prevalent modeling has been done for pairs of galaxies. The obvious reason for this is that the programming of more than two galaxies takes a huge amount of computing power, which simply does not yet exist.

There are several interesting and well-studied interacting pairs of galaxies. Models showing how these pairs formed their present features have been largely successful in that they reproduce the observed structures without much in the way of assumptions about mass, speed, angle of interaction, or halo (dark matter) mass. When structure can be reproduced using measurable and measured information, the modeling is considered successful. The most successful models have been used to further predict the unobserved (or, actually, the as-yet-unobserved).

Unlike stars, galaxies appear not to have formed in clusters, but to have clustered together over time. This clustering effect will be discussed in greater detail in chapter 10. What is most interesting about studying interacting galaxies is that while there are clearly some galaxies that formed from the beginning as elliptical galaxies, it appears that many formed as the result of a merge event between two galaxies. In fact, the conditions are most favorable for producing an elliptical galaxy from a merging pair, no matter what the morphological type of the galaxy before the merger.

It turns out that it is much less likely that a spiral galaxy will be the result of a galaxy merger than that an elliptical galaxy will be the result. To get anything other than an elliptical galaxy, you have to start out with two spirals whose disks are aligned to one another during the merger event. In addition, the two disks must be rotating so that when they merge, the material in each disk simply becomes a part of the resultant disk, so they would have to be rotating with the same speed to not disrupt the disk completely.

Knowing this, the fact that there are so many spirals (about half the galaxies in the universe) is a puzzle. Either the probabilities of the outcomes of the interactions that scientists have come to is incorrect, or most galaxies have not experienced a merge event. Of course, the latter is the most likely answer. As the universe expands, the space between galaxies increases, making it less and less likely that galaxies will interact, much less merge. In the present universe, we see very few merger events, but many interacting pairs and groups of galaxies. And, in the early universe we see far more interacting and even merging pairs of galaxies; however, even in the early universe, not all the galaxies are merging, or even interacting. The fraction of galaxies interacting in the early universe was much greater than in the present universe. Additionally, the fraction of interacting galaxies that are merging was also greater.

This is largely because galaxies were closer together in the early universe, so they interacted more. In addition to this, the objects that merged to become galaxies may have formed in regions of over-density. This means that galaxy seeds were already in clusters so that they would interact and merge and form present-day galaxies. This also explains why so many galaxies in the present universe are members of clusters or groups.

When galaxies interact or merge, the gas within them is used to form stars. Such bursts of star formation occur in the stellar history of every elliptical galaxy. One can use models to predict the color, spectral energy distribution, and gas content of galaxies, given a particular star formation history. Doing this, astronomers have concluded that every elliptical galaxy is an elliptical galaxy because near the beginning of its existence, it experienced a large burst of star formation. As such, a majority of stars in the galaxy, now, are near the ends of their evolutionary sequences. The galaxy appears red because most stars are red giants or red dwarfs.

The spectral energy distribution indicates that most of the energy coming from elliptical galaxies is at longer wavelengths, indicating that the stars in these galaxies are emitting low-energy photons, so they have low surface temperatures and are mostly on the right-hand side of the H-R Diagram, indicating that they are evolved (red giants) or very low mass (red dwarfs). Also, elliptical galaxies have little or no gas and dust in them. This is consistent with the proposed scenario of a burst of star formation. Such a burst would use up all available gas and dust in the formation of stars.

Since all elliptical galaxies fit this model, it is not inconsistent with the proposed hypothesis that all elliptical galaxies are the results of merger events that occurred early in the history of the universe. This is not proof that all elliptical galaxies formed this way, but recent observations of early universe are beginning to show evidence that supports this hypothesis.

ANDROMEDA AND THE MILKY WAY

Our own Milky Way is a member of an interacting pair of galaxies. There is some evidence that the Milky Way is involved in a gravitational interaction with its nearest large neighbor, Andromeda. Andromeda is the nearest large spiral galaxy to the Milky Way. It is located about 600 kpc away from the Milky Way.

Based on the measurements of the motions of both the Milky Way and Andromeda relative to one another, it appears that they are an interacting pair of galaxies. Currently, Andromeda and the Milky Way are careening toward one another at 300 km/s, over 210,000 miles per hour. Astronomers predict the collision will occur in about 3 billion years. The timing for the collision depends a lot on Andromeda's tangential motion. The above measurement of Andromeda's motion is measured using Doppler shift, which measures radial motion (towards or away). Andromeda is one of the small percentage of galaxies in the universe that is moving towards the Milky Way. Because of its proximity to the Milky Way, its motion toward us is significant. We cannot measure the tangential motion of Andromeda as it is too slow (based on the available data, any tangential motion Andromeda has must be less than one tenth of one arcsecond per century).

The most recent models predict that the collision of Andromeda with the Milky Way will result in an elliptical galaxy about one billion years after the collision begins. Several simulations of this predicted collision can be found on the Web. The prediction is made considering the initial speeds of the two galaxies as well as the mass of the two galaxies and the inferred dark matter halo mass. Recent observations put the dark matter halo masses of Andromeda and the Milky Way at roughly 10^{12} $M_\odot$ (solar masses) each. This is approximately equal to the mass of each galaxy, so that the dark matter halo accounts for about half the mass of each galaxy.

Astronomers predict that the Milky Way and Andromeda will merge during this collision. Most of the gas and dust within both galaxies, if not all of it, will be used to form a new generation of stars during the merging process. What will happen to Earth and our solar system? Probably, not much will change. It does depend on where the Sun is, in relation to gas clouds, during the collision, but most likely the Sun will be moved around, relative to the center of the Milky Way, during the merging process. The Sun could end up closer or farther from the galactic center.

BARRED SPIRAL GALAXIES

Barred spiral galaxies are spiral galaxies whose bulge is extended along one axis, making it look more like a bar than a sphere. The formation of such a feature both causes and is caused by a gravitational disturbance. To elongate the spheroidal bulge requires some kind of gravitational perturbation. While astronomers have not been able to single out a particular mechanism that accounts for all bar formation, it is clear that at least some bars are due to galaxy interactions. It is also clear that the formation of a bar at the center of a galaxy causes perturbations in the disk of that galaxy, which may cause inner or outer rings to form, as well as forcing a two-arm spiral pattern.

Bars and Arms

A bar at the center of a disk is very different from a spheroidal bulge at the center of a disk. First, it changes the look of the gravitational potential well. A spheroidal bulge has a gravitational potential well that look like a cone. However, a bar makes the point of the cone wider. To complicate things more, the bar rotates, and this rotation twists the wide point of the cone. So, large bars can greatly affect the shape of the gravitational potential well of a disk galaxy.

The gravitational potential well plays a significant role in the development of spiral density waves. It is the framework on which the spiral density waves are laid. Depending on how strongly the bar affects the underlying gravitational potential, a bar can drive spiral density waves. Conversely, a small or weak bar can be present in a disk that has no spiral density waves.

The stronger and larger the bar, the more grand design the spiral arm pattern becomes. Large, strong bars can also extend beyond the **radius of**

corotation. In disk galaxies with bars, there are two different motions: the motion of the bar feature and the motion of the disk. Just as the spiral density wave pattern motion is different from disk rotation, the bar feature and the disk motion are separate. The bar feature rotates as a solid body in this case, so the stars appear to move faster with increasing distance from the center point; the disk, however, rotates differentially, so the gas and dust and stars move at the same rate, no matter how far they are from the center. The radius of corotation is where the disk and bar feature are rotating at the same rate.

While the stars are orbiting the center of the galaxy, they are not orbiting in circular, or even elliptical, orbits. The orbits of the stars within a galaxy are more complex. They involve a lot of perturbations (due to the gravitational potential well as well as gravitational interactions with other stars orbiting the galaxy). Most stars appear to orbit the galaxy in a "wobbly" pattern due to their orbits having **epicycles.**

Generally, if the radius of corotation is close to the location of a **resonance** (when the difference between the speed of the stars and the pattern speed of the bar or the disk is equal to a whole number of epicycles), it marks the location for an inner ring feature. Galaxies can have inner rings, nuclear rings, and outer rings. This ring structure is usually an indication of resonances that occur within a galaxy. Galaxies with strong resonances often are home to spiral density waves.

Resonances can also occur within a bar. Stellar orbits within a bar can vary widely, causing different kinds of bars. These shapes may or may not indicate the presence of density waves or the strength or size of the bar. As a result of the differing possible stellar orbits, bars can be long and narrow or wide and boxy or even, peanut shaped.

The presence of a bar in a disk galaxy with an existing spiral density wave system can cause multiple symmetries. That is, there may be a two-arm pattern near the center that changes to a three-arm pattern after corotation. Many unexpected symmetries have been discovered in spiral galaxies with bars and without bars. Multiple symmetries can occur when two competing spiral density waves are present, which does not require a bar, per se.

While there are galaxies that have bars and are flocculent (without spiral density waves), the bars in these galaxies do not extend to corotation. Most galaxies with bars that extend to corotation have a grand design arm pattern. Dynamically, this can be explained by the gravitational potential of the galaxy being dominated by the bar causing a perturbation within the disk, which results in a spiral density wave. Observations so far support this hypothetical scenario. However, not every galaxy in the universe has been assigned a spiral arm class, or classified by its bulge shape. So, while astronomers seem to be in agreement that this explains what is currently observed, the explanation is contingent upon what is observed. If new observations are counter-indicative of this hypothesis, then it will be altered.

Bars and Interactions

Bars are not observed in every spiral galaxy in the universe; however, most spiral galaxies contain bars. This fact alone is indicative that bars must be transient, but long-lived features. That is, bars must be able to exist and then disappear. Also, bars must be able to appear where they were not before. One mechanism for the creation of bars is galaxy interactions. When galaxies interact, their gravitational potential wells are disturbed. A bar is a perturbation, or disturbance, within a gravitational potential well. Therefore, astronomers think it is possible that bars can be caused by galaxy interactions. This would explain the prevalence of bars in the universe, while allowing for the existence of a significant population of galaxies without bars.

There is some observational evidence to support this. Many barred galaxies show evidence of galaxy interactions. Some show evidence because they have nearby neighboring galaxies that may be currently interacting gravitationally. Some show evidence because they have had past epochs of rapid star formation. However, not every barred spiral galaxy has obvious evidence of galaxy interactions.

A possible mechanism for destroying and re-creating bars is **active galactic nuclei (AGN),** which are thought to be supermassive black holes found in the centers of some galaxies (both spiral and elliptical galaxies can have AGN). These supermassive black holes have **accretion disks** around them, which can extend to as far as several parsecs from the black hole. The accretion disk is just a disk of material that is being accreted by the black hole. Because of its mass (usually equal to or greater than 10^6 $M_\odot$), the black hole can attract matter to its accretion disk from great distances.

A bar feature works to fuel an AGN because it drives material towards the AGN. As it does so, it removes the bar feature by removing the gas clouds in the orbits that create the bar feature. It is mainly the inflow of gas, not stars that destroys the bar feature. Slowly, the inflow of gas causes the bar feature to bulge out, becoming more lens-like, and, eventually, creating a bulge. Since the AGN mainly consumes gas, it is this inflow that destroys the bar. However, once the bar is destroyed the inflow of gas stops, which shuts down the AGN, allowing the bar to re-form.

More detail about how AGN function will come in chapter 9; however, in a nutshell, the supermassive black holes that fuel AGN are thought to be the results of galaxy mergers. It is not clear whether supermassive black holes were formed before their host galaxies or vice versa. There are many objects found in the early universe that contain supermassive black holes and do not appear to have undergone any interactions or mergers, and yet almost every AGN in the present universe shows some evidence of a past interaction or merger event. It is not clear what role the supermassive black holes found at the cores of some galaxies play in galaxy evolution. Much of the answer to this puzzle comes from understanding how galaxies formed and evolved. This is the topic of chapter 10.

RECOMMENDED READINGS

Asimov, Isaac. *Asimov on Astronomy.* New York: Bonanza Books, 1988.

Bennett, Jeffrey D., Megan Donahue, Nicholas Schneider, and Mark Voit. *The Cosmic Perspective.* 5th ed. San Francisco: Benjamin Cummings, 2007.

DeGrasse Tyson, Neil, Charles Tsun-Chu Liu, and Robert Irion. *One Universe: At Home in the Cosmos.* Washington, DC: Joseph Henry Press, 1999.

Elmegreen, Debra Meloy. *Galaxies and Galactic Structure.* Englewood Cliffs, NJ: Prentice Hall, 1997.

Freedman, Roger, and William J. Kaufmann III. *Universe.* 8th ed. New York: W.H. Freeman Company, 2008.

Hawking, Stephen. *On the Shoulders of Giants.* Philadelphia: Running Press, 2002.

Ratay, Douglas L., *Multi-wavelength Observations of Barred, Flocculent Galaxies.* Available at http://etd.fcla.edu/UF/UFE0005401/ratay_d.pdf.

Rubin, Vera C. *Bright Galaxies, Dark Matters.* New York: AIP Press, 1996.

Seeds, Michael A. *Astronomy: The Solar System and Beyond.* 5th ed. Pacific Grove, CA: Brooks Cole, 2006.

9

Active Galaxies

Active galaxies really refers to a family of a variety of types of galaxies that have one thing in common, they have active galactic nuclei (AGN). An active galactic nucleus is a supermassive black hole located at the center of the galaxy, which is actively taking in gas through an accretion disk. The process of accreting matter through a disk causes x-ray, gamma ray, and synchrotron radio radiation, as well as, depending on the orientation of the disk relative to Earth, an extremely bright core (in the visible part of the electromagnetic spectrum), unusually narrow or broad emission lines, or jets.

SEYFERT GALAXIES

Seyfert galaxies are a class of galaxies named for their discoverer, Carl Seyfert. Carl Seyfert discovered this class of galaxies in 1943. They are described as a class of spiral galaxies whose nuclear spectra have unusual emission lines. Since the centers of spiral galaxies contain mostly stars and very little gas, the spectrum of a spiral center should be an absorption spectrum matching the spectral type of mostly evolved stars. However, Seyfert found that there were many spiral galaxies that had emission spectra, not absorption spectra. This type of spectrum indicates the presence of hot, ionized, gas. Even more interesting, this hot, ionized gas found in the cores of Seyfert galaxies is not associated with the kinds of stars that are usually associated with ionized gas. There were actually two different types of emission spectra that Seyfert observed, which is why there are two kinds of Seyfert galaxies. These subclasses were not identified until 1974, when higher resolution spectra were able to pick up these subtle features.

Seyfert Type 1

Seyfert 1 galaxies are galaxies whose nuclei have spectra with both broad and narrow emission lines. The broad lines are a result of the Doppler effect. Because the gas is moving very fast (usually thousands of kilometers per second), the wavelength of the light emitted is shifted. Since the gas in the nucleus of a Seyfert galaxy is moving in an accretion disk, sometimes it is moving towards Earth and sometimes it is moving away from Earth. When the gas is moving towards Earth, the light it emits is shifted to the blue end of the spectrum. When the gas is moving away from Earth, the light it emits is shifted to the red end of the spectrum. So, since we are observing the entire disk, some of the light is shifted to the blue end of the spectrum and some of the light is shifted to the red end of the spectrum. As a result, we see a line that is very wide extending from a much shorter wavelength than the emitted light to a much longer wavelength than the emitted light.

Also, the width of the emission lines indicates the presence of different densities of gas. The broader the emission lines, the denser the gas. This is because the broader lines are indicative of higher velocities for the particles within the gas that is emitting light. The faster moving gas, must be more dense in order to stick together at such high speeds. As a result of these observations of broad line emission spectra, astronomers have concluded that Seyfert 1 galaxies are spiral galaxies that have supermassive black holes at their cores that are accreting material from an accretion disk, which is emitting light as it rotates and drops material on the central supermassive black hole.

The spectra of Seyfert 1 galaxies indicate that in the nuclei of these galaxies there is both high-density and low-density gas emitting light. The high-density gas is in the accretion disk, because it is rotating. The low-density gas is not in the accretion disk, rather, it is in a **torus** (donut-shaped ring) of gas that surrounds the accretion disk. This is the gas that produces the narrow lines. This gas is not moving as fast as the denser gas, so the emission lines from this region are not as broad. The narrow-line producing gas is also thought to be associated with a ring of star formation that may be the result of the inflow of gas to the accretion disk to fuel the supermassive black hole. The inflow of gas to the accretion disk causes gas clouds to collide, which causes star formation, so surrounding the accretion disk is a ring of star formation. This ring of star formation is the source of the narrow emission lines.

Seyfert Type 2

Seyfert 2 galaxies do not exhibit the broad lines in their spectra. Rather, they appear to only have narrow emission lines at their cores. Since the **central engine** for both Seyfert 1 and Seyfert 2 galaxies is the same, astronomers think the reason that the broad emission lines are not seen in Seyfert

2 galaxies is because the accretion disk is obscured by the dusty torus (donut-shaped ring) the surrounds the accretion disk. This model is consistent with observations that indicate the presence of an accretion disk and black hole due to synchrotron radio radiation.

Synchrotron radio radiation is a particular type of radio radiation that is caused by the acceleration of electrons around helical magnetic field lines. That is, because of the accretion disk around the supermassive black hole, jets are formed that lie along the axis perpendicular to the accretion disk. These jets are essentially twisted magnetic field lines that extend outward. Along these twisted magnetic field lines, charged particles (including electrons) are accelerated along these lines. As the charged particles move along these magnetic field lines, they emit low-energy electromagnetic radiation—radio radiation.

It is called synchrotron radio radiation because it is the result of electrons that are moving at speeds near the speed of light in a circular pattern; the motion of the electrons is helical. Synchrotron radio radiation is distinguished from other radio radiation by its spectral energy distribution. A blackbody would show a peak at a shorter wavelength and decrease in intensity towards longer wavelengths. Synchrotron radiation peaks at longer wavelengths and decreases in intensity towards shorter wavelengths.

RADIO GALAXIES

Radio galaxies are galaxies that emit most of their electromagnetic radiation in the radio regime of the electromagnetic spectrum. Often, they have a visible jet, or two jets that are symmetrical. These jets are associated with large **radio lobes** that extend much farther than the galaxy itself. The majority of the radio radiation from these radio lobes is synchrotron radiation, indicating that the material in the lobes contains charged particles and a strong magnetic field.

In addition to the radio radiation, the lobes, jets, and cores of these galaxies emit a lot of x-ray radiation. The recent x-ray satellites have added much to our understanding of radio galaxies because of their ability to detect and map the locations of the x-ray radiation associated with these objects. X-ray radiation is indicative of a process that is capable of producing such high-energy radiation. The lobe and jet features are also indicative of the nature of the compact energy source at the cores of these galaxies. A supermassive black hole with an accretion disk that is actively accreting material can explain the existence of the jets, the radio lobes, and the x-ray radiation.

The mechanisms for creating these features are fairly straightforward, in the scheme of things. Material accreting onto the black hole by way of the accretion disk gets heated up to extremely high temperatures. This extremely hot material radiates x-ray radiation. Within the accretion disk, material is experiencing extremely strong forces. The gravitational pull of the supermassive black hole is tearing material apart because the strength of the force is so great that tidal forces are strong enough to tear apart atoms. So, as the material is accreted, it is broken down into its smallest pieces freeing electrons within the disk, which will move along established magnetic field lines creating synchrotron radio radiation.

Tidal Forces

Tidal forces are forces that are caused by a large difference between the strength of a force on the near and far sides of a solid object. For example, the gravitational force of Jupiter, at very close distances from Jupiter, decreases significantly with distance. That is, at the distance of the average orbit of Io, its nearest satellite, Jupiter pulls much more on the near side of Io than on the far side of Io. That difference is large enough to stretch Io and cause it to be extended along the line of the gravitational pull of Jupiter. As a result, Io is alternatively stretched and squished throughout its orbit of Jupiter. This keeps the core of Io molten and makes the surface of Io active and young in appearance. This is how tidal forces can work to tear apart solid objects. The more massive the gravitational source, the greater the difference in force strength at distances close to the object.

So, the tidal forces caused by being very near a supermassive black hole are strong enough to tear apart atoms. This is significant because, comparatively, the gravitational force is 38 orders of magnitude weaker than the nuclear force, which holds the nuclei of atoms together; however, supermassive black holes are unusual objects around which rare things do occur.

As the accreting material approaches the supermassive black hole, some of it crosses the event horizon of the black hole and becomes a part of the hole, but some of it is expelled from the black hole to conserve charge and angular momentum. Although the black hole is so massive and compact that its escape velocity for a large radius is greater than the speed of light, if material does not cross the event horizon, it can escape the gravitational pull of the black hole if it has enough energy. The ejected material is charged, and must have very high energy to escape the black hole. This charged material moving at relativistic speeds is responsible for both the x-ray and synchrotron radiation that is observed in the jet and lobe features around radio galaxies.

Event Horizon

A black hole is point in space known as a singularity. The object itself is dimensionless. A black hole is a point of mass and energy. It is said that even light cannot escape a black hole. In itself, this last statement is not entirely an accurate description of a black hole. The mass of a black hole is so compact that there is a relatively large area around a black hole where the escape velocity is greater than the speed of light. This area is called the **event horizon.**

The size of the event horizon of a black hole depends on its mass. A small black hole will have a small event horizon, but a massive black hole will have a larger event horizon. The event horizon describes the area around a black hole where the escape velocity exceeds the speed of light. Once an object crosses the event horizon, it cannot escape the gravity of the black hole and it is doomed to become part of the black hole.

Radio galaxies are mostly elliptical galaxies. Many of them are giant elliptical galaxies. Giant elliptical galaxies are located at the centers of large clusters of galaxies. Presumably, these giant elliptical galaxies are the results of multiple merger events. Sometimes giant elliptical galaxies are described as cannibal galaxies because they lie at the bottom of the deep gravitational potential well of the dense galaxy cluster in which it sits and all the other galaxies in the cluster are moving towards or interacting gravitationally with the giant elliptical galaxy.

QUASARS

The word "**quasar**" is short word for the actual name of this object: quasi-stellar radio source. When first discovered, it was not clear what the nature of these objects was. They appeared to be point-like sources of light, so they were called quasi-stellar objects, or QSOs. The reason they were not considered stars is because their spectra were redshifted as much as (or more than) the most distant galaxies, and yet they were much brighter than these distant galaxies. Since there were not very many of these objects known at first, and there were no models or theories that could explain how they could produce so much energy and still appear to be point sources.

As observations increased the number of quasars known, astronomers began to develop models for how so much energy could be produced in such a small object. Debate grew about whether the redshifts of quasars meant the same thing as the redshifts of galaxies. Perhaps there was another way to cause a redshift that wouldn't imply that these objects were so distant. The great distance of quasars presented a problem only in that if quasars were as far away as their redshifts indicated, they were the brightest objects in the universe—brighter than any known objects.

Eventually, a handful of quasars were found to be hosted by spiral galaxies. These observations were made with the newly repaired Hubble Space Telescope and with the extremely large ground-based telescopes. Now, astronomers can show by spectroscopic analysis and high resolution imagery that quasars are the bright nuclei of very distant galaxies.

At this point in time, it is commonly accepted that quasars are an early universe type of AGN. Conditions in the cores of these galaxies are different than those in the galaxies we see that are nearer to the Milky Way. In the cores of galaxies that host quasars, there is more gas than in the cores of nearby galaxies. If the mechanism for producing energy in quasars is the same as in nearby AGN, there is much more fuel for the fire, so to speak, in the quasar phenomenon than in either the radio galaxy or the Seyfert phenomenon. This is why quasars are so much brighter than their nearby AGN counterparts.

The other observable trait of quasars is the presence of superluminal **ejecta.** Several quasars have been observed to eject material (observed in the radio).

Over the years, astronomers have been able to observe the ejected material moving away from the quasar source. The rate at which these ejecta move away from the quasars is faster than the speed of light. Obviously, the material cannot really be moving faster than the speed of light, but simple calculations of the rate of speed of these objects does result in speeds greater than *c*. The effect that allows these ejecta to appear to move at **superluminal speeds** is a relativistic effect called **Doppler boosting.**

Doppler Boosting

The universe is expanding at a rate of about 72 km/s/Mpc. That means that an object that is 1 megaparsec distant is increasing in distance by 72 km every second. That is not very much, since a megaparsec is 3.086×10^{19} km. So in one second an object that is 1 megaparsec distant goes from being 30,860,000,000,000,000,000 km away to being 30,860,000,000,000,000,072 km away. At large distances, this can become significant. The distance at which most quasars are found is greater than 1,000 Mpc, so the change in length of the linear distance is 72,000 km every second. An object at this distance that is moving close to the speed of light (299,792 km/s) will appear to be moving faster than the speed of light because of the rate of expansion of the space between the observer and the object. This object will appear to have superluminal speed.

Doppler boosting is an artifact of the expansion of the universe. When viewing objects that are at great distances from Earth, the motion of the expansion of the universe is the bulk of the motion observed. For objects that are moving very fast (near the speed of light) the expansion of the universe adds motion that can make the object appear to have superluminal motion. However, no object can move at a speed that exceeds the speed of light, so this is just an apparent motion.

BL LAC OBJECTS AND BLAZARS

Like variable stars, as astronomers began observing galaxies, they identified classes that had similar behaviors. The first object in the class became the archetype and bore the name for the class. Such is the case for BL Lacertae—a galaxy located in the constellation Lacertae at the location of the star known as BL. It turns out that it is one of many galaxies that behave this way. The group is known as **blazars.** Perhaps this particular type of AGN maintains its variable star-like name because it is the one type that is characterized by rapid and large energy variations. At first, this galaxy was thought to be a variable star, hence the name.

After studying this object for a long time, the fluctuations in energy appeared to be irregular and constant. This object and others similar to it are characterized by featureless optical spectra, core radio sources, and jets exhibiting superluminal motion. They are high redshift (very distant) objects that are very bright, like quasars. Yet, they do not appear to have significant amounts of gas or dust near or around them. As such, astronomers have concluded that **BL Lac objects** are hosted by elliptical galaxies.

Radio and x-ray observations of these objects show that they are sources of high-energy photons and synchrotron radiation. The observed super-luminal motions are indicative of jets that appear to exhibit superluminal motion because of Doppler boosting, as in quasars. Yet, these objects are different from quasars in that their spectra do not indicate the presence of gas or dust.

THE CENTRAL ENGINE

In the core of each of these types of AGN lies a supermassive black hole with an accretion disk and a torus of dust surrounding it. This is known as the central engine. The characteristics of this central engine are empiri-cally derived. That is, observations have led to these conclusions. Astrono-mers have observed similar energy amounts and energy distributions from these AGN.

Each source is contained in a very small space or exhibits variations over very short time scales (minutes or seconds). The small size of the central engine is derived from this fact. Since nothing can change faster than the speed of light, the variations in brightness observed in these different types of AGN limit the possible size of the central energy source. Variations in intensity that occur in minutes imply sizes not much bigger than our solar system.

The enormous amounts of energy observed coming from these extremely small sources makes these objects extremely interesting. Astronomers have developed a model of a system that can explain all the observations of AGN so far. Different AGN appear different due to their environments and orien-tations. The current Unified Model explains every observation to date. New observations are used to test the model. If something is observed that cannot be explained by the model, it will be altered until it can explain the observed phenomenon. It is approaching the theory state, but there is still much more to learn about supermassive black holes and their behavior before the Uni-fied Model can graduate to theory.

THE UNIFIED MODEL

The Unified Model was first proposed in the 1980s. Astronomers began to notice that these four classes of galaxies had much in common. They all exhibited high-energy x-ray radiation, variability over short timescales, syn-chrotron radio radiation and radio jets, lobes or ejecta. So the energy source of these objects must be massive and small.

Given the amounts of energy released and small timescales over which variations are observed, the size of the energy source must be similar to the size of the solar system. The only type of object that could be so small and so massive is a supermassive black hole. The energy production would be due to material being accreted onto the black hole by way of an accretion disk that surrounds the supermassive black hole.

From observing these four different kinds of AGN, and noting their many and varied features, astronomers have concluded that these objects have this central engine in common. In addition to the supermassive black hole and surrounding accretion disk, there is a torus of gas and dust. This torus has the effect of blocking the view of the disk from certain angles, directing the jets of material and associated radio lobes, being the source of fuel for bursts of

Diagram of Active Galactic Nucleus Model

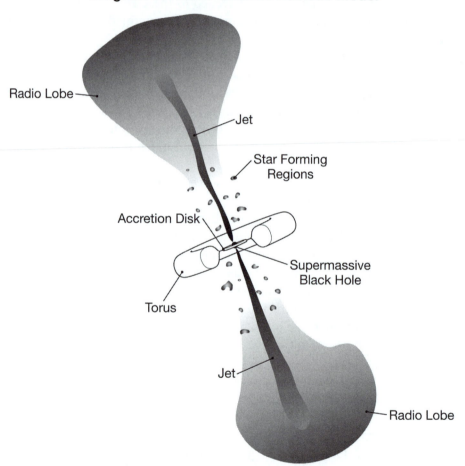

Figure 9.1 The schematic above depicts the current model of the central engine that powers all active galactic nuclei (AGN). At the center is the supermassive black hole. Surrounding that is an accretion disk of material. Around that is a torus of dust and gas. Perpendicular to the accretion disk, jets and radio lobes can be observed. Depending on the angle at which the lobes are viewed, they may both be seen, only one, or none. [Jeff Dixon]

star formation and serving as a mechanism for re-starting the central engine when the accretion disk runs out of material.

Astronomers have created this model based on all the observations for all the different types of active galaxies observed in the universe. The model was first conceived to explain the difference between Seyfert 1 and Seyfert 2 galaxies. Astronomers who observed radio galaxies, quasars, and BL Lac objects began to see that this proposed energy source could explain their observations as well. Interestingly, the connection of this energy source for all active galaxies further enhanced the attempt to put together galaxy evolution. Since quasars and BL Lac objects are high-redshift, early universe objects, and since Seyferts and radio galaxies are low-redshift, recent-universe objects, the fact that one energy source model can explain all active galaxies means that such an energy source is an important connection between early-universe objects and objects in the current epoch. Galaxy evolution and how active galaxies have helped piece that together is the topic the next and final chapter.

RECOMMENDED READINGS

Asimov, Isaac. *Asimov on Astronomy.* New York: Bonanza Books, 1988.

Bennett, Jeffrey D., Megan Donahue, Nicholas Schneider, and Mark Voit. *The Cosmic Perspective.* 5th ed. San Francisco: Benjamin Cummings, 2007.

DeGrasse Tyson, Neil, Charles Tsun-Chu Liu, and Robert Irion. *One Universe: At Home in the Cosmos.* Washington, DC: Joseph Henry Press, 1999.

Freedman, Roger, and William J. Kaufmann III. *Universe.* 8th ed. New York: W.H. Freeman Company, 2008.

Hawking, Stephen. *On the Shoulders of Giants.* Philadelphia: Running Press, 2002.

Seeds, Michael A. *Astronomy: The Solar System and Beyond.* 5th ed. Pacific Grove, CA: Brooks Cole, 2006.

10

Galaxy Evolution

Like stellar evolution, the evolution of galaxies is not something that humans can observe since it takes much longer than a human lifetime for a galaxy to change. The fact that galaxies change is evident from observing galaxies at different epochs in the universe. Looking at all the snapshots of galaxies over time, astronomers have pieced together a galaxy evolution scenario. Additionally, they have considered what is known about the composition of the universe (the amount of **cold dark matter** and hot dark matter and dark energy) to further test and constrain this model.

In this chapter, we will explore the various environments of galaxies, the idea of **look-back time,** and what is known about galaxy evolution. Finally, we will examine the most recent galaxy evolution models and describe what data they are based on and what future observations will help to constrain this model and how they will constrain it.

GROUPS

As discussed in chapter 8, galaxies are often found in the company of one another. The Milky Way, for example, is part of a group of galaxies known as the Local Group. This grouping of galaxies is a loose affiliation of a small number (about 50) of galaxies. A group of galaxies is gravitationally bound; that is, each galaxy's motion is due its interaction with the gravitational force of every other galaxy in the group. The average size of a group is about 1 to 2 Mpc in diameter. Most, if not all, of the galaxies within 1 Mpc of the Milky Way are part of the Local Group.

Because we live in the Milky Way, there is a large portion of the sky that is not visible. This is because the Milky Way is blocking our view. For that reason, we cannot say that astronomers have identified every galaxy in the Local Group. Recently two large galaxies were discovered that are members of the Local Group. Both galaxies are located exactly in the region of the sky that is obscured by the dusty disk of the Milky Way. Both are fairly large galaxies compared to the others in the Local Group. So, it would not be unexpected if smaller galaxies were found that are, by chance, located in the region of the sky that is blocked from our view by the dusty disk of the Milky Way.

Besides the Local Group, there are other groups of galaxies, but most of them are much smaller than the Local Group. Many are also compact groups of galaxies (discussed in more detail in chapter 8). Most galaxy groups are named for their central member, such as the M81 group.

CLUSTERS

Clusters of galaxies are different from groups primarily in the number of galaxies there must be to make the group a cluster. Clusters contain more than 50 and less than about 1,000 galaxies. Moreover, they are slightly larger in size. They can be from 2 to 5 Mpc in diameter. Clusters can appear to be organized or not. Most clusters have one or two large elliptical galaxies at their centers. These large galaxies are clearly the result of many galaxy mergers. Some clusters, however, have no central galaxy or several (more than two) large elliptical galaxies that are spread throughout the cluster, indicating several cores or no core at all.

Additionally, clusters exhibit evidence of dark matter. Clusters of galaxies are gravitationally bound, but their members have speeds that are fast enough for them to escape the gravitational potential well of the cluster. Just like with the gas and stars in galaxies, this extra speed is indicative of unseen mass that is required for these speeds to be below the escape velocity of the cluster. In addition to these clusters having large amounts of dark matter, they also emit a large amount of x-ray radiation due to a hot **intracluster medium.** This hot intracluster medium is just hot gas that has escaped from the galaxies, perhaps during merger events in the history of the cluster. Because the intracluster medium is very hot (usually 10 to 100 million K), the atoms in the gas emit mostly high energy x-ray photons.

While the number of galaxies in a cluster can vary along with the size, the density of galaxies in the cluster has a much smaller range. The density of the galaxies (the number per volume) is what classifies a cluster as rich or poor. The dividing line is a count of the number of galaxies above a certain brightness level within a particular radius from the center. A rich cluster will have more than 30 galaxies that are no less than two magnitudes fainter than the third brightest galaxy within a radius of 1.5 Mpc from the center. If the count of galaxies of that brightness within that radius is less than 30, the cluster is considered poor. Groups are also considered a type of poor cluster.

SUPERCLUSTERS

In some regions of space, astronomers have discovered large strings of clusters that seem to be a clustering of clusters. These structures are known as **superclusters.** There are not many of these known to date, but they appear to be a common phenomenon. Earlier in our understanding of the universe, the clustering of clusters was thought to indicate a special place in the universe, towards which everything, ultimately, gravitated. It was even thought that something dubbed The Great Attractor might be the cause for the clustering of galaxy clusters. While astronomers are sure that there is more than one such location in the universe, it is still not well understood why clusters should cluster at all.

Beyond superclusters are **supercluster complexes,** which appear to be even larger structures like "The Great Wall" of galaxies found in the high redshift survey in the late 1980s. These structures are even more of a mystery to astronomers. Perhaps they indicate the existence of an underlying structure to the universe that is, as yet, undiscovered.

REDSHIFT

To understand how all these pieces fit into the current understanding of the universe and galaxy evolution, the concept of redshift is a necessary piece of information. Astronomers can measure the motion of things in the universe using **redshift.** This phenomenon is an aspect of a well-known phenomenon observed on Earth known as Doppler shift.

Doppler shift is the phenomenon of waves changing wavelength due to the motion of a source or observer. Doppler shift occurs when a sound source, like a train, is moving towards an observer. In this case, the sound the train is making is shifted to a higher pitch as the train approaches. When the source is moving away from the observer, the sound is shifted to a longer wavelength, resulting in a deeper pitch as the train recedes. The same phenomenon can occur if the observer is moving and not the source. For example, as an observer approaches a loud carnival, the sounds get higher and higher in pitch until the observer passes the carnival and the sounds shift to lower and lower pitches.

In astronomy, we consider that the observer is stationary and the source is moving, although it really does not matter which one is moving—the source or the observer. The waves in astronomy are light waves, not sound waves, so the shift is in color, not pitch. When an object is moving toward the observer, the light emitted is shifted towards shorter and shorter wavelengths, so the light appears bluer than it should. This is a **blueshift.** Conversely, when an object is moving away from an observer, the light emitted is shifted to longer and longer wavelengths, so the light appears redder than it should. This is a redshift.

Because the universe is expanding, the space between Earth and every galaxy in the universe is being stretched. So most galaxies are redshifted, due to the Doppler Effect. Some galaxies are not redshifted because they are moving towards Earth faster than the universe is expanding the distance between Earth and the galaxy, so their net motion is still towards Earth. This can really only occur for galaxies that are relatively close to Earth. This is because it does not take a very great distance for the rate of expansion of the universe to exceed normal galaxy speeds.

The more the light is redshifted, the faster the object is moving away. Because of the expansion of the universe, the faster a galaxy is moving away from Earth, the farther away it must be. Therefore high redshift is the same as great distance. Redshift is measured in kilometers per second or fraction of the speed of light. That is, an object that is moving away from Earth at a speed equal to half the speed of light has a redshift of 1.5×10^5 km/s or 0.5.

Even though nothing in the universe can actually go faster than the speed of light, some galaxies have redshifts as high as eight, meaning their light is shifted so much that their speed away is eight times the speed of light. Since the motion of a galaxy due to the expansion of the universe is only apparent (in that the galaxy is not actually moving, rather the space between the observer and the galaxy is increasing due to the universe expanding), galaxies can have redshifts that are greater than the speed of light.

LOOK-BACK TIME

Look-back time is a strange phenomenon of the vast distances we deal with in astronomy combined with the finite speed of light. Besides the Moon, there are no objects close enough to Earth that the light we observe from them is nearly instantaneous. The Sun is eight light minutes away, which means that the Sun is so far away that if it could stop emitting light, it would take eight minutes until we were able to tell! The nearest star, besides the Sun, is over four light-years away. That means that the light we observe when we look at this star is light that left the star more than four years ago. So, if we want to see what this star looks like right this instant, we have to wait more than four years.

Imagine you have three friends, Abe, Beth, and Carl, who lived 5, 10, and 20 light years away from you, respectively. Each decides to send a picture of themselves to you on their 21st birthday. You would receive Abe's picture first, but by the time you got it, Abe would be 26 years old. Beth's picture would arrive when she was 31, and Carl's picture would arrive when he was 41. So, who looks youngest for their age? The friend who lives farthest away is the one who appears much younger than he is.

Galaxies are millions of light years from Earth. This means that the light we see when we look at other galaxies is light that left that galaxy millions (or more) of years ago. To see what a distant galaxy looks like at this instant, we have to wait millions of years. This may seem like a big problem—that we never see anything as it is now. But, astronomers have thought of this as a way to see the past. That is, when we observe distant galaxies, we can see how galaxies looked millions (or billions) of years ago. If we put them on a distance-time scale, we can try to piece together how galaxies evolve. So, observing distant galaxies can be like watching a movie of how galaxies evolve, if you can piece the parts together correctly.

This idea of observing distant objects to look at how things were in the past is called look-back time. Since more distant objects are younger (seen as they were longer ago), astronomers can observe what galaxies looked like when they first formed, if they can see objects that are distant enough.

ACTIVE GALACTIC NUCLEI, THEN AND NOW

Applying this to active galactic nuclei (AGN), an obvious evolutionary sequence is apparent. Recall, from chapter 9, that Seyfert galaxies are AGN found in the local universe (not much look-back time), and quasars are AGN found at high redshift. Both types of AGN are found in mostly spiral galaxies. If the central engine model is correct, Seyfert galaxies may be older versions of quasars. Radio galaxies and BL Lac objects share the same relationship since they are both housed in primarily elliptical galaxies.

Another similarity between the Seyfert—quasar and radio galaxy—BL Lac pairs exists. In both cases the high redshift object is more energetic and more concentrated. This implies that AGN may start out very energetic and may lose energy over time. Interestingly, the number of AGN in the local universe and at high redshift are not significantly different. That is, there are just as many young AGN as there are old AGN in the universe. Also, there appears to be a peak in high redshift AGN at about redshift of 1 to 2.

The most recent models suggest that this is due to an epoch of galaxy-galaxy mergers. Such merger events could trigger the formation of super-massive black holes as well as help to fuel them by driving all remaining gas and dust to the core of the galaxy where it can become part of the accretion disk or gas and dust torus.

THE EVOLUTION PICTURE

Astronomers studied stars in different environments to determine the evolution story of stars. From these observations, they learned that stars go through phases where different processes produce the energy they emit. Their intrinsic properties change in predictable ways. Knowing all these changes and processes, astronomers were able to understand our own star, the Sun, better. Finally, understanding stars led to understanding our Galaxy. The Milky Way contains both young clusters and old clusters of stars. These different populations indicate different epochs of star formation.

Looking at the Milky Way, astronomers can identify and study the stellar populations and learn that the spheroidal components (bulge and halo) formed stars before the disk did. Looking at other galaxies, stellar population models can help to deconvolve the history of star formation in those galaxies. Most spiral or disk galaxies show a similar star formation history to that of the Milky Way—old spheroidal components and a younger disk. This modeling can explain isolated or nearly isolated galaxies, but what about galaxies in clusters?

The question of galaxy formation and evolution involves a lot of cosmology. Actually, the constituents available to form galaxies and the stars within them have been around since the epoch of recombination (when the universe cooled to a temperature of 10,000 K). But when did galaxies form and how did they form? Part of this question is answered by determining whether galaxies were built from smaller pieces or broken off of larger ones. This is a question of top-down or bottom-up evolution.

Stars appear to form in a top-down scenario. A cloud of gas and dust breaks apart into stars. Galaxies, on the other hand, appear to form a bottom-up scenario. That is, first small proto-galaxies formed, and they merged and coalesced to form the galaxies we know today.

The clearest picture of galaxy evolution comes from the extraordinary images produced by space telescopes in the last few decades. The advantage of being in space is more than just not having to look through Earth's blurring atmosphere. There is also the advantage of being able to observe in a particular direction for longer than a night on Earth will allow. Since a space telescope can always point away from the Sun, it never really experiences a time when it cannot observe the sky. Instead, there are regions of the sky that cannot be observed because the Sun or Moon or Earth are in the way.

With the advent of space-based astronomy, someone had the ingenious idea to spend some time observing the same part of the sky for a longer time than anyone had ever done before—11 days. The telescope did not actually make the observation for 11 straight days. Modern astronomers use electronic light detectors, similar to the digital cameras found in most cell phones today, to capture images of the sky; the detectors used in astronomy, however, can collect light for very long time periods (minutes to hours). To capture these revolutionary images, the space telescope pointed to the same part of the sky and took images at least 342 separate times. Because the images are digital, astronomers can add them together, so that faint objects appear brighter. The total exposure time of all the images when added together is a little longer than 11 days!

The Hubble Deep Field

In December 1995, the newly launched and recently repaired Hubble Space Telescope pointed to a part of the sky that contained no bright stars and no bright galaxies to speak of. The telescope imaged this part of the sky for such a long time that the faintest objects in the image have an apparent magnitude of almost 32. This is 16 magnitudes fainter than the human eye can see (that is about 10 trillion times fainter).

The Hubble Deep Field was the first deep image that showed astronomers what lurks on the edge of the visible universe. Astronomers saw, for the first time, the building blocks of galaxies, and were able to confirm that galaxies most likely formed from the joining together of smaller pieces, rather than the breaking apart of larger pieces. The smaller pieces appear to be unstructured and blue in color, relative to their more modern cousins. We know the small, blue, fuzzy objects found in the Hubble Deep Field are proto-galaxies because they are far away, which means we are seeing them as they were a long time ago.

The Hubble Ultra-Deep Field

The Hubble Deep Field was such an important datum for astronomy that it was decided to do another Hubble Deep Field in the southern part of the sky. This was an important experiment to make sure that what was observed in the original Hubble Deep Field was not unique, but represented what is seen in every direction. The southern Hubble Deep Field demonstrated that there was nothing special about the Hubble Deep Field. The southern Hubble Deep Field was so important that astronomers thought it would be a great idea to extend the depth of the Hubble Deep Field. This is how the Hubble Ultra-Deep Field was conceived.

The Hubble Ultra-Deep Field is a new image taken with two different cameras on the Hubble Space Telescope. The image took 1 million seconds of exposure time and is of another relatively empty part of the night sky. This image was taken simultaneously with two different instruments—one an imager and one a spectrograph. This way, each object's spectrum is also available. For the thousands of objects observed in the Hubble Ultra-Deep Field, most have been identified. Some of the objects in this image are the youngest objects ever seen. These objects are so young that the objects identified as building blocks of galaxies in the Hubble Deep Field are now understood to be an intermediate stage between the actual building blocks and the galaxies of the present epoch.

THE EVOLUTION OF GALAXIES

So, what do astronomers now know about the evolution of galaxies? First, it is clear that galaxies form from smaller pieces. This is what observations of the most distant objects reveal. This conclusion has implications for understanding the universe since it implies that most of the dark matter in the universe is cold, rather than hot. This means that there should be more objects like MACHOs (massive compact halo objects) than like massive neutrinos (weakly interacting massive particles)—WIMPs. Cold dark matter is simply dark matter that is slow-moving. (Dark matter is discussed in much greater detail in chapter 7.)

The space-based telescope deep-sky images have also revealed that the formation of galaxies early on was by merging the proto-galaxies observed in the early universe. This means that galaxy formation is bottom-up process. This further implies that the formation of clusters and groups of galaxies is a product of this same process. The current model for cluster formation begins with the small over-densities observed in the cosmic microwave background radiation.

These seeds are where there may have been slightly more proto-galaxies than in other parts of the universe. Those proto-galaxies attracted other proto-galaxies. Eventually the proto-galaxies merged to form galaxies. In the areas where the seeds were, there were more galaxies than in other areas. These galaxies attracted other galaxies. Over time, the gravitational attraction of the seeds to other galaxies created groups and clusters of galaxies. The galaxies and proto-galaxies that formed these groups and clusters came from nearby and far away, leaving empty space behind them.

Some astronomers think it is possible that within the great voids in the universe (where there are few galaxies) there may be proto-galaxies that never merged (for lack of neighbors) and have evolved in isolation. Studies of so-called isolated galaxies attempt to explore galaxy evolution in undisturbed environments. Since almost everything we call a galaxy must have been through at least one merger in its existence, isolated galaxies offer a unique laboratory for studying galaxy evolution.

Another topic in the realm of galaxy evolution is the presence of bars. Bars appear to be transient structures that are fairly long-lived. Some dynamic models suggest that bars may vary in strength over time. If bars are a mechanism to fuel a central supermassive black hole, the idea of varying strength could be consistent with the idea of turning on and turning off a central engine. But, not all galaxies with bars appear to house an active nucleus.

The difference between elliptical and spiral or disk galaxies appears to be star formation history. Star formation history may be a way to read the galactic evolution story for an individual galaxy. That is, each galaxy may have its own history that can be told by looking at star formation history and determining how stars have formed over time (whether looking at an early burst of star formation with no recent activity or at nearly constant star formation rates that are fairly low with perhaps some peaks of star formation that indicate strong interactions or merger events).

Within disk galaxies there are some with prominent spiral density waves and others that have no evidence of spiral density waves at all. Disk galaxies appear to vary widely in arm class. Very little is understood about why spiral density waves are present in some galaxies while not in others. The presence of spiral density waves, too, may be a transient feature that is related to mergers or galaxy-galaxy interactions.

The theme that should be evident by now is that galaxy mergers are a very important part of galaxy evolution. Galaxy mergers may drive galaxy evolution. Studies of galaxies in the process of merging or interacting violently will lead to a better understanding of how this process changes galaxies. Studies of galaxies that only rarely interact with other galaxies may lead to a better understanding of how galaxies can evolve without the aid of merger events. The evolution that occurs without galaxy-galaxy interactions is probably driven by stellar evolution and may be a slower process than the changes that can occur during merger events and other types of galaxy interactions.

Stellar evolution takes millions to billions of years to occur. It took only about a century to deduce the processes of stellar evolution through observations of clusters of stars (see chapter 3 for more on star clusters). Galaxy evolution takes at least tens of billions of years. It has only be a century since astronomers have known that galaxies other than the Milky Way exist, so astronomers may be close to understanding galaxy evolution, or it may be another century from now before this tale is unraveled. Certainly, our understanding of this process has made great gains thanks to the deep images of the space-based telescopes. With new technology comes greater opportunity for deeper understanding.

RECOMMENDED READINGS

Asimov, Isaac. *Asimov on Astronomy.* New York: Bonanza Books, 1988.

Bennett, Jeffrey D., Megan Donahue, Nicholas Schneider, and Mark Voit. *The Cosmic Perspective.* 5th ed. San Francisco: Benjamin Cummings, 2007.

DeGrasse Tyson, Neil, Charles Tsun-Chu Liu, and Robert Irion. *One Universe: At Home in the Cosmos.* Washington, DC: Joseph Henry Press, 1999.

Freedman, Roger, and William J. Kaufmann III. *Universe.* 8th ed. New York: W.H. Freeman Company, 2008.

Hawking, Stephen. *On the Shoulders of Giants.* Philadelphia: Running Press, 2002.

Seeds, Michael A. *Astronomy: The Solar System and Beyond.* 5th ed. Pacific Grove, CA: Brooks Cole, 2006.

Glossary

absolute magnitude (M). A measure of the energy emitted by a star that uses a logarithmic scale.

absorption spectrum. A type of spectrum that includes almost all wavelengths of visible light (those excluded are determined by the chemical composition of the absorbing medium). It looks like a rainbow with dark lines in it.

accretion disk. A disk of material that is orbiting a black hole or other mass source (like a star); the material in the disk will become a part of the black hole by spiraling toward it.

active galactic nucleus (AGN). A type of galaxy core that emits a huge amount of high-energy electromagnetic radiation as well as synchrotron radiation.

apparent magnitude (m). A measure of the energy received from a star that uses a logarithmic scale.

arcminute. An angular measurement equal to 1/60th of a degree of arc.

arcsecond. An angular measurement equal to 1/60th of an arcminute or 1/3600th of a degree of arc.

asterism. A group of stars that are part of one or more constellations; such groupings are not recognized as constellations, but are often used to locate objects in the sky.

astrometry. The use of accurate measures of positions of stars to determine the rates of motion or positions of stars or other celestial objects.

astronomical unit (AU). A linear distance equal to 1.5×10^{11} m. The average distance between Earth and the Sun.

asymptotic branch. The stage of red giant phase achieved only by stars massive enough to fuse carbon in their cores.

band. A part of the electromagnetic spectrum sampled using a filter that allows only a small range of electromagnetic radiation to pass through it and blocks all the rest.

barred spiral galaxies. A disk-dominated galaxy with a central bulge that is elongated or peanut-shaped.

barycenter. The center of mass of a system of multiple objects.

beta decay or proton decay. The process whereby a neutron decays into a proton, during this process an electron and an electronic anti-neutrino are emitted.

BL Lac Objects. Objects like the galaxy known as BL Lac. High-redshift AGN housed primarily in elliptical galaxies.

black dwarf. A theoretical star that would be the final stage of a white dwarf when the star cools to a temperature of absolute zero and emits no light.

blackbody. A theoretical object that absorbs all radiation incident upon it and radiates all wavelengths of light. A hot solid or hot dense gas.

blackbody spectrum. A plot of the intensity vs. wavelength (or frequency) of the radiation from a blackbody. The range of intensities per wavelength or frequency of electromagnetic radiation emanating from a hot solid or hot dense gas.

blazars. High-redshift astronomical objects that house active galactic nuclei (AGNs) and are associated with elliptical galaxies. (Also known as BL Lac objects.)

blueshift. The shift of light toward the blue (shorter wavelength) end of the spectrum due to motion towards a source.

Bohr model. A model of what an atom looks like that is useful for explaining the phenomena of emission and absorption spectra; the model describes an atom as a nucleus with electrons that exist only on specific energy states or orbits.

brightness. The amount of energy received from a light-emitting source.

bulge. The spheroidal component of a disk-dominated galaxy located at the center; a bulge that is elongated or peanut-shaped is called a bar.

central engine. The power source of an active galactic nucleus. A supermassive black hole with an accretion disk, torus of dust, radio lobes, jets and star forming regions at the core of an active galaxy.

Cepheid variable. A type of pulsating variable star that exhibits a period-luminosity law so that knowing the period of variation of the star allows astronomers to determine its luminosity, thereby making it possible for astronomers to calculate its distance from Earth.

Cerenkov radiation (pronounced "Chair-in-koff"). Electromagnetic radiation that is a result of high-energy, nearly massless particles, called neutrinos, interacting with normal matter.

clusters of galaxies. Groups of more than 50 and up to 1,000 galaxies; clusters are 2 to 3 Megaparsecs in diameter.

cold dark matter. Slow-moving matter that does not radiate any electromagnetic radiation, but has mass.

constellation. A group of stars defined by the International Astronomical Union that define a region in the sky.

contact binary. A binary system of stars where the two stars are touching one another.

continuous spectrum. A type of spectrum that includes all wavelengths of visible light. This looks like a rainbow.

convective transfer. The transfer of energy by convection or the movement of material that contains thermal energy; the material is heated, making it less dense and causing it to rise; as it rises, it radiates heat and becomes more dense, sinking.

dark matter. Matter that does not radiate any electromagnetic radiation; as yet, it is not known what type of matter this might be.

dark nebulae. Clouds of dust that block the light of stars and gas behind them.

declination (Dec). The celestial sphere equivalent of lines of latitude on Earth measured in degrees; the line of zero degrees is a projection of Earth's equator on the sky and is called the celestial equator.

differential rotation. A mode of rotation about a center where every part moves at a different angular rate around the center, but with the same linear rate.

diffraction. The spreading out of light due to the deformation of a wave as it passes through an aperture.

disk. The flattened part of a galaxy that contains most of the gas and dust. A flattened shape of gas and dust (as in accretion disk).

disk galaxies. Galaxies that have a flat disk-like component that dominates the shape of the galaxy.

dissociate. To break a molecular bond, as in: molecular hydrogen dissociates into atomic hydrogen due to high-energy photons.

Doppler boosting. A relativistic effect that causes ejecta moving away from distant active galactic nuclei to appear to be moving at speeds greater than the speed of light.

Doppler Effect. A phenomenon of waves that occurs when either the source or the observer is moving; motion away results in longer wavelengths observed than emitted, while motion towards results in shorter wavelengths observed than emitted.

dwarf spheroidal galaxies. Very small galaxies that appear to be roundish in shape and contain little or no gas or dust.

dwarf star. A type of star that is small; can be either red dwarf (a type of main sequence star), brown dwarf (a failed star), white dwarf (a type of hot stellar remnant that does not fuse nuclei to produce energy), or a black dwarf (a cooled off white dwarf that no longer emits any light).

ejecta. Materials that are ejected by an active galactic nucleus or a star.

electromagnetic radiation. All radiation that is in the form of photons; radiation caused by the acceleration of charged particles, and is the interaction of electric and magnetic fields.

electromagnetic spectrum. The range of wavelengths and/or energy of electromagnetic radiation.

elliptical galaxies. Spheroidal galaxies or galaxies that are roundish in shape and contain little or no gas and dust.

emission spectrum. A type of spectrum that includes only a small fraction of the wavelengths of visible light (those included are determined by the chemical composition of the emitted medium).

epicycles. Loops in an otherwise circular or elliptical orbit.

event horizon. The point of no return around a black hole; beyond the event horizon approaching the black hole, the escape velocity exceeds the speed of light.

extinction. The process of attenuating light; light that is either absorbed or scattered is considered extinguished.

flocculent galaxies. Galaxies that have no dominant spiral structure, but have bright, fleecy spurs of star formation throughout their disks.

gamma ray. The most energetic, shortest wavelength form of electromagnetic radiation.

giant elliptical galaxies. The largest of all galaxies, these appear to be roundish in shape and contain little or no gas and dust.

giant star. A type of star that is very large and fuses helium into carbon.

globular cluster. A dense, spheroidal cluster of 10,000 to 1,000,000 stars, usually found in the halo of a galaxy.

grand design galaxies. Galaxies that have strong spiral arm structure, usually two bright arms that extend from the bulge to the outer limits of the disk.

gravitational lens. A massive object that bends the space-time continuum enough to change the path of light.

ground state. The lowest energy an electron can have; the closest (to the nucleus) orbital an electron can be in around its nucleus.

groups of galaxies. Groups of less than 50, but more than 3 galaxies; groups are 1 to 2 Megaparsecs in diameter.

halo. The spheroidal component of a disk-dominated galaxy that extends beyond the visible disk component and contains the globular clusters and is dark matter dominated.

horizontal branch. The second stage of the red giant phase denoting stable helium fusion in the core.

H-R Diagram or Hertzsprung-Russell Diagram. A graph of stars with temperature or spectral type on the x-axis and luminosity or absolute magnitude on the y-axis; this diagram can be used to infer the evolutionary sequence of stars.

infrared. The region of the electromagnetic spectrum that includes wavelengths just longer than visible light.

instability strip. A region that runs through the red giant, horizontal and asymptotic branches where stars are unstable and, due to their oscillations, become variable stars, varying their brightnesses over periods of hours to years.

interference. The interactions of waves with one another; this interaction can be additive (constructive) or subtractive (destructive) causing light and dark spots.

interstellar medium. The material between the stars within a galaxy—usually gas (hot, warm, and cold) and dust.

intracluster medium. The material found between galaxies within a cluster.

ionize. To lose an electron, as in: atomic hydrogen is ionized by ultraviolet radiation.

Keplerian motion. A mode of rotation about a center where the objects that are closer move faster and objects that are farther move slower.

look-back time. The phenomenon that objects that are far away are seen as they were long ago.

low surface brightness galaxies. Galaxies that have a lower density of stars in their disks; as such, these galaxies are not very bright compared to the night sky, so they are difficult to detect.

luminosity. A measure of the amount of energy per second given off by a light-emitting source (usually the energy is emitted by a star or some other astronomical object).

main sequence star. A type of star that fuses hydrogen into helium.

metallicity. A measure of the amount of elements heavier than helium in any celestial object.

microwave radiation. The region of the electromagnetic spectrum between infrared (the region that includes wavelengths just longer than visible light) and radio (the least energetic and longest wavelength electromagnetic radiation).

Milky Way. The name for the galaxy in which we live.

molecular clouds. Large conglomerations of molecular gas (usually hydrogen or carbon dioxide).

nebula. A cloud-like celestial object that may be gas heated by stars, dust obscuring starlight, dust reflecting starlight, the expanding outer shells of a dying star, or distant galaxy or group of stars; plural: *nebulae*.

neutrino. A nearly massless, structureless subatomic particle; a type of lepton.

neutron star. A type of stellar remnant that gives off light; its composition is thought to be neutrons since its density is so great that electrons and protons would not be able to be separate particles; these stars do not fuse nuclei to produce energy; the energy emitted by these stars is thermal energy.

open cluster. A loose association of stars, usually found in the plane of a galaxy.

parallax. A way to measure distance to relatively nearby objects using the fact that observations from opposite ends of a baseline result in different positions of nearby objects relative to more distant objects. The change in apparent position of a nearby object relative to more distant objects; measured in tenths or hundredths of an arcsecond for stars.

parsec (pc). A unit of distance equal to 3.6×10^8 m or 3.26 light years.

photoionization. The process of ionization (losing an electron) by way of interaction with electromagnetic radiation (photons).

photon. A particle of light; photons are massless, but contain energy; the amount of energy a photon has determines the classification of the electromagnetic radiation.

planetary nebula. A cloud-like formation that is caused by the outer layers of intermediate mass star, like our Sun, being ejected at the end of its final stages of the red giant phase of its evolution.

polycyclic aromatic hydrocarbons (PAHs). Long chains of hydrogen and carbon atoms that form the smallest dust grains.

pre-main sequence stars. Stars that are not yet stable enough to fuse hydrogen in their cores, but are progressing towards that end.

proper motion. The motion of a star relative to Earth that is *not* due to Earth's rotation around the Sun.

quasar. Quasi-stellar radio source; a high-redshift active galactic nucleus associated with spiral galaxies.

radiative transfer. The transfer of energy by radiation via photons.

radio lobes. Large regions of synchrotron radio radiation associated with galaxies that house active galactic nuclei.

radio radiation. The region of the electromagnetic spectrum that includes the least energetic and longest wavelength electromagnetic radiation.

radius of corotation. The distance from the center of a disk-dominated galaxy at which the stars move at the same rate as the spiral pattern.

red giant. The helium fusion phase of stellar evolution. In this phase, stars may also fuse heavier elements, but hydrogen fusion only occurs in a thin shell around the core.

red giant branch. The first stage of the red giant phase for stars evolving off the main sequence denoting shell hydrogen fusion, but not core helium fusion yet.

redshift. The shift in wavelength of light toward the red (longer wavelength) end of the spectrum due to motion of a source away from an observer or motion of an observer away from a source.

reflection nebulae. Clouds of gas and dust that scatter the light of stars in front of them towards an observer in the same direction.

refraction. The "bending" of light due to its speed changing as it passes through different media; different wavelengths of light experience different amounts of changes in speed and so, experience different amounts of "bending"; red changes direction the least, while blue changes direction more.

resonances. A phenomenon of coincidence of orbits between at least two orbiting masses that results in the amplification of an oscillation.

right ascension (RA). The celestial sphere equivalent of lines of longitude on Earth; RA is measured in hours, not degrees; the line of zero hours is one that goes through the constellation Aries.

sidereal time. Time measured relative to the apparent motion of the stars; a sidereal day of 24 sidereal hours is 23 hours 56 minutes of "real" time, so a sidereal second is slightly shorter than a "real" second.

solar luminosity ($L_\odot$). A unit of luminosity that is equal to the luminosity of the Sun: about 3.83×10^{33} erg/sec.

solar mass ($M_\odot$). A unit of mass that is equal to the mass of the Sun: about 1.99×10^{30} kg.

solar radius ($R_\odot$). A unit of distance that is equal to the radius of the Sun: about 6.96×10^8 m.

solid-body rotation. A mode of rotation where every part has the same angular rate around the center, but different linear rates.

spectral type, (also spectral class). A classification of a star's spectrum that can be used to identify its surface temperature.

spheroidal galaxies. Galaxies that are roundish in shape and whose constituents exhibit tri-axial motion (motion in all three spatial dimensions).

spin-flip transition. An energy transition that involves the flipping of a subatomic particle that has the quantum property of spin (a proton or an electron).

spiral density waves. The phenomenon of over-dense areas of interstellar material that causes spiral arm patterns in some disk galaxies.

spiral galaxies. Galaxies that are disk-dominated and usually exhibit some spiral structure within the disk, commonly known as spiral arms; the Milky Way is a spiral galaxy.

star-forming regions. Areas within a galaxy where very young stars are present; usually identified by ionized hydrogen gas.

starbursting galaxies. Galaxies that are undergoing a burst of star formation—usually indicated by an excess of ionized hydrogen emissions.

Stefan-Boltzmann Law. A mathematical relationship describing the dependence of luminosity on temperature and radius of the source.

supercluster complexes. The largest structures in the universe made of superclusters.

superclusters. Large structures made of clusters of galaxies.

supergiant. A type of star that is extremely large; this type of star fuses elements heavier than hydrogen into elements heavier than helium in its core to maintain equilibrium.

superluminal speed. A speed greater than the speed of light (2.99 x 10^5 km/s).

supermassive black hole. A singularity usually located at the center of a galaxy that contains a mass greater than about 1 million times the mass of the Sun.

supernova. An explosive event that marks the end of a massive star's existence; this occurs when thermonuclear fusion can no longer occur in the core of the star so that its outer layers collapse on its core.

synchrotron radiation. Electromagnetic radiation emitted by charged particles traveling at relativistic speeds (close to the speed of light).

T Tauri. A type of variable star that is associated with Sun-like stars in their pre-main sequence stage of evolution.

terrestrial. Anything that is found on Earth or due to Earth's atmosphere is characterized as terrestrial (as opposed to solar, or interstellar, for example).

thermonuclear fusion. The fusion of light-weight elements to form heavier elements occurring under conditions of extremely high temperatures and extremely high densities.

torus. A doughnut shape; usually this shape is associated with the clouds of dust and gas surrounding an accretion disk in the central engine of an active galactic nucleus.

triaxial. Something that uses all three spatial dimensions; as in, triaxial motion.

triple alpha process. A process for fusing helium into carbon using three steps and three alpha particles (a.k.a. helium nuclei).

turn-off point. The point on a color-magnitude diagram of a cluster where the main sequence turns off to the red giant branch.

ultra-compact dwarf galaxies. Extremely small dense groups of stars, gas, and dust considered to be some of the smallest and faintest galaxies in the universe.

ultraviolet. The region of the electromagnetic spectrum that includes wavelengths just shorter than visible light.

x-ray. The region of the electromagnetic spectrum between gamma rays (the most energetic and shortest wavelength) and ultraviolet (the region that includes wavelengths just shorter than visible light).

Bibliography

Aguilar, David. *Planets, Stars, and Galaxies: A Visual Encyclopedia of Our Universe.* Washington, DC: National Geographic Children's Books, 2007. An excellent volume for middle school and high school students.

Asimov, Isaac. *Asimov on Astronomy.* New York: Bonanza Books, 1988.

Bennett, Jeffrey D., Megan Donahue, Nicholas Schneider, and Mark Voit. *The Cosmic Perspective.* 5th ed. San Francisco: Benjamin Cummings, 2007.

Binzel, Richard P., series ed. *University of Arizona Space Science Series.* Tucson: University of Arizona Press, 1979.

The University of Arizona Press's long-running Space Science Series publishes cutting-edge research on planetary sciences. This series is relatively technical, aimed at graduate school–level students and designed as a general reference for professionals. The series is currently 30 volumes in total, but the books in the series relevant to the topics discussed in this volume are listed below. The series is periodically updated, so, for example, *Protostars and Planets V* supersedes *Protostars and Planets IV*; however, earlier versions of books can still provide interesting and useful information and so are also listed below.

Planets, Stars and Nebulae Studied with Photopolarimetry. Gehrels, T. 1974.

Protostars and Planets III. Levy, Eugene, and Jonathan I. Lunine, with the assistance of Mildred Shapley Matthews, and Mary L. Guerrieri. 1993.

Protostars and Planets IV. Mannings, Vince, A. P. Boss, and S. S. Russell. 2000.

Protostars and Planets V. Reipurth, Bo, David Jewitt, and Klaus Keil. 2007.

Chaisson, Eric, and Steve McMillan. *Astronomy Today: Stars and Galaxies.* Vol. II. 6th ed. Menlo Park, CA: Benjamin Cummings, 2007.

DeGrasse Tyson, Neil, Charles Tsun-Chu Liu, and Robert Irion. *One Universe: At Home in the Cosmos.* Washington, DC: Joseph Henry Press, 1999.

Elmegreen, Debra Meloy. *Galaxies and Galactic Structure.* Englewood Cliffs, NJ: Prentice Hall, 1997.

Freedman, Roger, and William J. Kaufmann III. *Universe.* 8th ed. New York: W.H. Freeman Company, 2008.

Hawking, Stephen. *On the Shoulders of Giants.* Philadelphia: Running Press, 2002.

Jastrow, Robert. *Red Giants and White Dwarfs.* New York: W.W. Norton and Company, 1990.

Ratay, Douglas L., *Multi-wavelength Observations of Barred, Flocculent Galaxies.* Available at: http://etd.fcla.edu/UF/UFE0005401/ratay_d.pdf.

Rubin, Vera C. *Bright Galaxies, Dark Matters.* New York: AIP Press, 1996.

Seeds, Michael A. *Astronomy: The Solar System and Beyond.* 5th ed. Pacific Grove, CA: Brooks Cole, 2006.

———. *Stars and Galaxies.* 6th ed. Pacific Grove, CA: Brooks Cole, 2007.

WEB SITES

Astronomical Objects Brightness Table: http://www.vaughns-1-pagers.com/science/star-magnitude.htm

Biography of Annie Jump Cannon: http://www.sdsc.edu/ScienceWomen/cannon.html.

Biography of Cecilia Payne-Gaposchkin: Astronomer and Astrophysicist: http://www.harvardsquarelibrary.org/unitarians/payne2.html.

Biography of M. Saha: http://banglapedia.search.com.bd/HT/S_0022.htm.

Biography of M. Saha: http://www.calcuttaweb.com/people/msaha.shtml.

Biography of Niels Bohr: http://nobelprize.org/nobel_prizes/physics/laureates/1922/bohr-bio.html.

Galaxies—World Book at NASA: http://www.nasa.gov/worldbook/galaxy_worldbook.html.

Mr. Sunspot's Record Book—Listing of Astronomical Records: http://eo.nso.edu/MrSunspot/records.

Niels Bohr's Open Letter to the United Nation, 1950: http://www.columbia.edu/~ah297/un-esa/ws1999-letter-bohr.html.

Occam's Razor: http://math.ucr.edu/home/baez/physics/General/occam.html.

Relative Brightness Table: http://www.stargazing.net/David/constel/magnitude.html

Spitzer Space Telescope: http://www.spitzer.caltech.edu/features/articles/20050627.shtml.

Stars Nearest Earth: http://www.essex1.com/people/speer/stars.html.

Stars—World Book at NASA: http://www.nasa.gov/worldbook/star_worldbook.html.

Index

About the Author

Lauren V. Jones completed her A.B. in both physics and astronomy at Vassar College, where she worked with Cindy Schwarz and Deborah Elmegreen. She earned her first M.S. in physics, with specialization in astronomy, at Moscow State University, where she worked with Anatoly Vladimirovich Zasov on the source of far infrared radiation in galaxies. Jones then returned to the United States to work with William Keel at the University of Alabama on her second M.S. in physics with specialization in astronomy. Her thesis incorporated more recent data and deconvolution techniques to analyze the contributions of different stellar populations to the far infrared emissions observed. She next completed an M.S. in astronomy and a Ph.D., working with Richard Elston, at the University of Florida. For five years after completing her Ph.D., Dr. Jones taught at different institutions of higher education. Becoming interested in science education research while working with Timothy Slater and Edward Prather at the University of Arizona, she took a position with the Ohio Department of Education.